I0408538

1

High-Resolution Protein Structure Determination: A Text-Based Informative Book

Rickbed Nandi

Preface

Welcome to the fascinating world of high-resolution protein structures! In this book, we will embark on a journey into the very heart of life's building blocks, proteins, and the incredible techniques that scientists use to uncover their hidden secrets.

Proteins are like the workers in the construction site of your body. They perform countless vital tasks, from digesting your food to defending you against diseases. Understanding how proteins work is like having a magic key to unlock the mysteries of life itself.

But there's a catch – proteins are incredibly tiny, much smaller than anything you can see with your eyes. So how do scientists study them? That's where the amazing world of high-resolution structural biology comes in.

In this book, we'll explore how scientists use tools like X-ray crystallography, NMR spectroscopy, and cryo-electron microscopy (cryo-EM) to take pictures of proteins at the tiniest level imaginable. It's like having a super-powered microscope that can see things a million times smaller than a grain of sand!

We'll journey through the chapters, each one unravelling a different aspect of this exciting field. From the basics of how these techniques work to the incredible predictions for the future, you'll learn about the heroes and heroines of structural biology who've dedicated their lives to solving the puzzles of protein structures.

You don't need to be a scientist to enjoy this book. We've written it in simple language so that anyone with a curious mind can join us on this adventure. Whether you're a student, a teacher, a science enthusiast, or just someone intrigued by the wonders of the microscopic world, there's something here for you.

So, turn the page, and let's dive into the mesmerizing world of high-resolution protein structures. By the time you finish this book, you'll have a whole new appreciation for the tiny heroes that make life possible and the incredible tools we use to discover their secrets. Get ready for a thrilling ride through the hidden universe of biology!

Rickbed Nandi

Dedication

This book is dedicated to students, instructors, and our cherished general audience who possess a basic understanding of the topic. Your curiosity, passion for learning, and unwavering support have fuelled our enthusiasm for sharing the wonders of high-resolution protein structures.

May the knowledge within these pages inspire you, empower you, and kindle a profound appreciation for the intricate world of structural biology. Your quest for understanding is the driving force behind the pursuit of scientific knowledge, and for that, we dedicate this book to you.

With gratitude,

Rickbed Nandi

Contents

Chapter 1: Introduction to Protein Structure Determination

1.1 Overview of the importance of protein structure

Proteins are the workhorses of life, performing a myriad of essential functions within biological systems. The three-dimensional structure of a protein is intricately linked to its function, and understanding protein structure is fundamental to unravelling the mysteries of life itself. In this subsection, we will explore the profound significance of protein structure by delving into its role in biology, medicine, and biotechnology, supported by relevant examples and data.

Protein Structure as the Foundation of Function

The primary structure of a protein, which refers to the linear sequence of amino acids, ultimately dictates its higher-order, three-dimensional structure. This intricate folding process leads to the formation of specific shapes and surfaces that are crucial for the protein's function.

Example 1: Enzymatic Activity

Enzymes, the catalysts of biological reactions, exemplify the intimate connection between structure and function. For instance, the enzyme lysozyme, found in tears and saliva, has a specific three-dimensional structure that allows it to cleave the cell walls of certain bacteria. This structural precision is crucial for its antibacterial function. A slight alteration in its structure could render it ineffective.

Example 2: Oxygen Transport

Haemoglobin, the protein responsible for transporting oxygen in our blood, is a prime example of structure-function relationships. Haemoglobin's four subunits fold together into a globular structure that can bind and release oxygen with remarkable precision. Alterations in its structure, as seen in sickle cell anaemia, disrupt its function, leading to severe health consequences.

Insights from Structural Biology

Structural biology, the scientific discipline focused on elucidating protein structures, has yielded invaluable insights into the inner workings of cells, organisms, and even diseases. One of the primary techniques for determining protein structures is X-ray crystallography.

Data: Impact of X-ray Crystallography

According to the Protein Data Bank (PDB), a repository of experimentally determined protein structures, there were over 185,000 structures deposited. This rich dataset has provided researchers with a wealth of information about protein structures across various organisms. Such data have been pivotal in understanding the mechanisms of diseases, drug design, and basic cellular processes.

Example 3: Drug Development

Many drugs target specific proteins in the body, and knowledge of the three-dimensional structure of these proteins is crucial for rational drug design. For instance, the anti-cancer drug Imatinib, which revolutionized the treatment of chronic myeloid leukaemia, was developed based on the crystal structure of the BCR-ABL kinase, a protein implicated in the disease.

Protein Structure and Disease

The role of protein structure extends beyond normal cellular function to the realm of diseases. Mutations or misfolding of proteins can lead to various disorders, underscoring the importance of understanding and characterizing protein structures for medical purposes.

Data: Protein Misfolding Diseases

Protein misfolding diseases, such as Alzheimer's, Parkinson's, and Huntington's diseases, are characterized by the abnormal aggregation of misfolded proteins. These diseases pose a significant burden on society. For instance, Alzheimer's disease affects approximately 5.8 million people in the United States alone, and its prevalence is expected to rise with an aging population.

Example 4: Alzheimer's Disease

The structure of amyloid-beta, a protein implicated in Alzheimer's disease, is critical for understanding the disease mechanism. Misfolded amyloid-beta forms plaques in the brain, leading to neuronal damage. Structural studies have shed light on the conformational changes that occur in amyloid-beta during disease progression, offering potential targets for therapeutic intervention.

Structural Genomics and Beyond

The advent of high-throughput structural genomics initiatives has accelerated the pace of protein structure determination. These projects aim to determine the structures of a vast number of proteins, providing a treasure trove of data for various applications.

Data: Structural Genomics Initiatives

The Structural Genomics Consortium (SGC) had determined the structures of thousands of proteins, contributing significantly to our understanding of the human proteome. Such initiatives have implications not only in understanding basic biology but also in drug discovery.

Example 5: Targeting Undruggable Proteins

Structural genomics efforts have focused on elucidating the structures of proteins previously deemed "undruggable" due to their complex functions or lack of known binding sites. These structures provide opportunities for designing novel therapeutics.

Biotechnological Applications

Protein engineering and biotechnology heavily rely on a detailed understanding of protein structures. The ability to modify and design proteins has led to advancements in various fields.

Example 6: Biopharmaceuticals

The biopharmaceutical industry has seen a surge in the development of protein-based drugs, such as monoclonal antibodies and recombinant proteins. Precise knowledge of protein structures is critical for ensuring the safety and efficacy of these therapeutics.

Example 7: Enzyme Engineering

In industrial applications, enzymes are often engineered for improved catalytic activity or stability. Structural insights guide these engineering efforts, resulting in enzymes optimized for processes like biofuel production or pharmaceutical synthesis.

Protein structure is the linchpin of biology, underpinning the functions of living organisms, the mechanisms of diseases, and the advancements of biotechnology. The vast body of knowledge accumulated through structural biology has had a profound impact on medicine, drug development, and our overall understanding of life at the molecular level. As technology continues to advance, we can anticipate even more transformative discoveries arising from the study of protein structures.

1.2 Historical development of protein structure determination techniques

The quest to decipher the three-dimensional structures of proteins has been a pivotal endeavour in the history of biochemistry and molecular biology. The determination of protein structures has not only provided fundamental insights into the functional mechanisms of life but has also paved the way for breakthroughs in drug discovery and biotechnology. This subsection delves into the historical development of protein structure determination techniques, tracing the evolution of methods from early biochemical insights to the high-resolution techniques of today.

Early Insights into Protein Structures: Theoretical Speculation and Chemical Approaches

The journey of unravelling protein structures begins in the late 19th century when scientists first speculated about the physical nature of proteins. In 1878, Swedish chemist Johannes Rydberg proposed that proteins might be composed of linear chains of amino acids, and this idea laid the foundation for the concept of the polypeptide chain. However, experimental techniques of the time could not provide direct evidence for this hypothesis.

In the early 20th century, X-ray crystallography emerged as a powerful tool for studying the atomic structures of crystalline materials. While initially applied to inorganic compounds, it soon became evident that this technique held immense potential for the study of biological macromolecules, including proteins. Max von Laue's pioneering X-ray diffraction experiments on crystals in 1912 marked a significant milestone, demonstrating that X-rays could scatter when they encountered regularly spaced arrays of atoms.

The Advent of X-ray Crystallography in Structural Biology

The true breakthrough in protein structure determination came in the 1930s when British biophysicist William Astbury employed X-ray diffraction to study fibres such as wool and silk, which are predominantly composed of proteins. Astbury's work revealed the first glimpses of the periodic structures within proteins. In 1934, he coined the term "secondary structure" to describe the repeating structural patterns observed in these fibres.

However, the direct application of X-ray crystallography to proteins faced significant challenges due to the difficulty of obtaining well-ordered protein crystals. This challenge was overcome by the work of American biochemist Linus Pauling, who, in the late 1940s, proposed the alpha-helix and beta-sheet as fundamental structural motifs in proteins based on hydrogen bonding patterns. His pioneering work provided crucial insights into the principles of protein folding.

The Myoglobin Breakthrough: The First Protein Structure Determination

The first successful determination of a protein's three-dimensional structure occurred in 1957 when British biochemist John Kendrew and his colleagues unveiled the structure of myoglobin, an oxygen-binding protein found in muscle tissue.

The myoglobin structure, determined by X-ray crystallography, represented a monumental achievement and was awarded the Nobel Prize in Chemistry in 1962. This landmark discovery opened the door to a new era in structural biology.

The myoglobin structure revealed that proteins indeed possess intricate and highly specific three-dimensional shapes. Myoglobin's structure featured a heme group at its core, which coordinated oxygen binding, confirming Pauling's ideas about the importance of non-covalent interactions in protein structure and function.

X-ray Crystallography Comes of Age: The Era of High-Resolution Structures

Following the myoglobin breakthrough, X-ray crystallography became the gold standard for protein structure determination. The technique continued to advance, enabling scientists to determine the structures of increasingly complex proteins. In 1960, the structure of haemoglobin, another oxygen-binding protein, was elucidated by Kendrew's team, further solidifying the role of X-ray crystallography in structural biology.

Throughout the 1960s and 1970s, X-ray crystallography contributed to the determination of numerous protein structures, providing insights into enzymes, antibodies, and viral proteins. Notable examples include the structures of lysozyme and ribonuclease, both of which earned their discoverers Nobel Prizes in Chemistry.

The NMR Spectroscopy Revolution

While X-ray crystallography was making significant strides, another technique, nuclear magnetic resonance (NMR) spectroscopy, was quietly developing as an alternative approach to protein structure determination. NMR spectroscopy had its roots in the study of nuclear magnetic resonance in the 1940s, but it was not until the 1960s that it was applied to biological macromolecules.

In 1951, American physicists Felix Bloch and Edward Purcell were awarded the Nobel Prize in Physics for their development of NMR, which allowed researchers to probe the magnetic properties of atomic nuclei. Over the following decades, NMR spectroscopy evolved into a powerful tool for studying protein structures in solution, offering insights into protein dynamics and interactions that were difficult to obtain through crystallography.

The 1980s: NMR and Recombinant DNA Technology Take Centre Stage

The 1980s witnessed significant advancements in both X-ray crystallography and NMR spectroscopy, with each technique contributing to a growing catalogue of protein structures. Recombinant DNA technology, which emerged in the 1970s, enabled the production of large quantities of pure proteins, revolutionizing structural biology by providing researchers with ample material for analysis.

In 1984, the structure of the protein, human insulin, was determined by X-ray crystallography. This landmark achievement not only had profound implications for diabetes treatment but also highlighted the synergy between biotechnology and structural biology.

The Post-Genomic Era: High-Throughput Structural Biology

The completion of the Human Genome Project in the early 2000s marked a watershed moment in biology, generating an explosion of sequence data. This genomic information provided a treasure trove of potential drug targets and proteins of interest, fueling the need for high-throughput structural biology methods. During this period, structural genomics initiatives emerged, aiming to determine the structures of a vast number of proteins rapidly. These large-scale efforts combined X-ray crystallography, NMR spectroscopy, and other biophysical techniques, leveraging automation and robotics to increase throughput.

The Cryo-EM Revolution: From Low-Resolution to Near-Atomic Resolution

In recent years, cryo-electron microscopy (cryo-EM) has revolutionized structural biology. While cryo-EM had been used for decades to study large macromolecular complexes and viruses, it was limited in its ability to resolve atomic details of protein structures.

However, advancements in detector technology, image processing algorithms, and sample preparation techniques in the early 21st century transformed cryo-EM into a high-resolution method. Researchers can now determine protein structures at near-atomic resolution using single-particle analysis, making cryo-EM a powerful complement to X-ray crystallography and NMR spectroscopy.

The historical development of protein structure determination techniques is a testament to the relentless pursuit of scientific knowledge. From the early speculations of protein structure to the high-resolution structures of today, the field has evolved significantly, driven by innovation, collaboration, and interdisciplinary approaches. The combination of X-ray crystallography, NMR spectroscopy, and cryo-EM has revolutionized our understanding of proteins, providing critical insights into the molecular machinery of life and opening up new avenues for drug discovery and biotechnology.

1.3 Basic principles of high-resolution structure determination

High-resolution protein structure determination is a fundamental pursuit in the field of structural biology. It aims to reveal the precise three-dimensional arrangement of atoms within a protein molecule. Understanding these structures at atomic resolution is crucial for unravelling the mechanisms of biological processes and for applications ranging from drug discovery to enzyme engineering. This subsection explores the fundamental principles that underlie high-resolution protein structure determination and highlights the key techniques employed in the process.

X-ray Crystallography: Illuminating Crystals with X-rays

One of the pioneering techniques in high-resolution structure determination is X-ray crystallography. This method relies on the interaction between X-rays and the electrons in the atoms of a crystal to generate diffraction patterns that reveal the arrangement of atoms in a protein. The basic principles of X-ray crystallography include:

Bragg's Law: Bragg's Law, formulated by William Henry Bragg and William Lawrence Bragg in 1913, is a fundamental equation in X-ray crystallography. It relates the angle of incidence (θ) of X-rays on a crystal lattice to the wavelength of the X-rays (λ) and the spacing between atomic planes (d). Mathematically, it can be expressed as:

$2 * d * \sin(\theta) = n * \lambda$

Where:

- **d** is the distance between crystal planes.
- **θ** is the angle of incidence.
- **n** is an integer representing the order of the diffraction peak.

Bragg's Law is essential for understanding how X-rays interact with crystals and how to interpret the resulting diffraction patterns.

Fourier Transform: Another critical concept in X-ray crystallography is the Fourier transform. The diffraction pattern generated by X-rays is a complex interference pattern of scattered waves. Applying a Fourier transform to this pattern converts it into an electron density map, which reveals the distribution of electrons in the crystal. The electron density map can then be used to determine the positions of atoms within the crystal.

Data Collection and Processing: In practice, X-ray crystallography involves collecting a series of diffraction images from a protein crystal. These images are processed to extract the amplitudes and phases of the diffracted waves, which are used to calculate the electron density map. Data processing algorithms, such as those in the CCP4 suite or Phenix software, play a crucial role in this step.

NMR Spectroscopy: Probing Molecular Interactions in Solution

Nuclear Magnetic Resonance (NMR) spectroscopy is another powerful technique for high-resolution structure determination, particularly for proteins in solution. NMR relies on the magnetic properties of certain atomic nuclei, such as hydrogen and carbon, to obtain structural information. Key principles of NMR include:

Nuclear Spins: Nuclei with an odd number of protons and/or neutrons possess a nuclear spin, which gives rise to a magnetic moment. In NMR, the behaviour of these nuclear spins in a magnetic field is exploited to gather structural information.

Chemical Shifts: Different atomic nuclei resonate at slightly different frequencies in an NMR experiment due to their local chemical environment. These variations in resonance frequency are known as chemical shifts and provide valuable information about a molecule's structure.

Spin-Spin Coupling: Spin-spin coupling occurs when nuclear spins interact with each other through chemical bonds. These interactions manifest as multiplet patterns in NMR spectra and reveal information about the connectivity of atoms in a molecule.

Resonance Assignment: To determine a protein's structure using NMR, it is essential to assign specific NMR signals to individual atoms in the molecule. This is achieved through a process known as resonance assignment, which is often facilitated by multidimensional NMR experiments.

NOE (Nuclear Overhauser Effect): NOE is a critical parameter in NMR structure determination. It results from the dipole-dipole interactions between nuclear spins and provides distance restraints between atoms in close proximity, aiding in the calculation of three-dimensional protein structures.

Cryo-Electron Microscopy (Cryo-EM): Freezing Biological Macromolecules in Action

Cryo-Electron Microscopy (Cryo-EM) has gained prominence as a high-resolution structural biology technique, allowing researchers to study protein structures without the need for crystallization. Key principles of Cryo-EM include:

Single-Particle Analysis: Cryo-EM involves imaging individual protein particles embedded in a thin layer of vitrified ice. These 2D images are then used to reconstruct 3D structures using advanced computational methods. Single-particle analysis is particularly valuable for flexible or heterogeneous samples.

Electron Microscopy Imaging: Cryo-EM employs a transmission electron microscope (TEM) to project electrons through the specimen. The interactions between the electrons and the sample generate contrast, which is captured in the form of micrographs.

Image Processing and 3D Reconstruction: Cryo-EM data undergo extensive image processing and 3D reconstruction to convert 2D micrographs into high-resolution 3D density maps. This process involves aligning and averaging thousands of individual particle images.

Resolution Enhancement: Advances in Cryo-EM technology, such as direct electron detectors and improved data processing algorithms, have significantly enhanced the achievable resolution. Modern Cryo-EM can achieve near-atomic resolution. These fundamental principles of high-resolution structure determination underpin the three major techniques: X-ray crystallography, NMR spectroscopy, and Cryo-EM. While each technique has its strengths and limitations, they collectively contribute to our understanding of protein structures, enabling breakthroughs in fields as diverse as drug discovery, enzymology, and structural genomics. As technology and methodologies continue to evolve, the boundaries of what can be resolved at high resolution are constantly being pushed, opening new possibilities for unravelling the intricacies of biological macromolecules.

Chapter 2: X-ray Crystallography

2.1 Fundamentals of X-ray diffraction

X-ray diffraction is a fundamental technique in structural biology that plays a pivotal role in elucidating the atomic-level structures of biological macromolecules, particularly proteins and nucleic acids. This subsection explores the basics of X-ray diffraction, its historical significance, the principles underlying the method, and its application in protein structure determination.

Historical Perspective

The roots of X-ray diffraction can be traced back to the early 20th century when Max von Laue, Walter Friedrich, and Paul Knipping first demonstrated that X-rays could be diffracted by crystals. This breakthrough laid the foundation for the development of X-ray crystallography as a powerful tool for determining the three-dimensional atomic structures of crystalline materials, including biological macromolecules.

One of the most significant milestones in the history of X-ray crystallography was the work of Sir William Henry Bragg and his son, Sir William Lawrence Bragg. In 1912, they formulated

Bragg's law, a fundamental equation that describes the relationship between the wavelength of X-rays, the angle of incidence, and the spacing between crystal lattice planes. Bragg's law revolutionized X-ray crystallography, enabling scientists to interpret X-ray diffraction patterns and determine the atomic arrangements within crystals.

Principles of X-ray Diffraction

X-ray diffraction is based on the principles of wave interference and the wave-like nature of X-rays. When X-rays, which are electromagnetic waves with very short wavelengths (typically on the order of 0.1 nanometres), are incident on a crystal, they interact with the regularly spaced atoms in the crystal lattice. This interaction results in the scattering of X-rays in various directions.

Bragg's law, which governs X-ray diffraction, can be expressed as:

$2d\sin(\theta)=n\lambda$

Where:

- d is the distance between crystal lattice planes.
- θ is the angle of incidence.
- n is an integer corresponding to the order of the diffraction peak.
- λ is the wavelength of the incident X-rays.

According to Bragg's law, constructive interference occurs when the path difference between X-rays scattered from adjacent crystal planes is a whole number multiple of the X-ray wavelength. This results in the formation of diffraction peaks on a detector, and the angles at which these peaks are observed provide information about the crystal lattice spacing (d) and, ultimately, the three-dimensional arrangement of atoms within the crystal.

Experimental Setup

X-ray diffraction experiments require a highly controlled environment to achieve accurate results. The key components of an X-ray diffraction setup include:

X-ray Source: X-ray generators, such as rotating anode generators or synchrotrons, produce intense beams of X-rays with well-defined wavelengths.

Crystal Sample: The biological macromolecule of interest is crystallized into a highly ordered crystal lattice. The quality of the crystal directly impacts the quality of diffraction data.

Collimation: X-ray beams are collimated to form a focused beam that impinges on the crystal sample.

Detector: A detector records the diffraction pattern produced when X-rays interact with the crystal. Modern detectors, such as charge-coupled devices (CCDs) or pixel detectors, provide high-resolution and rapid data acquisition.

Goniometer: A goniometer allows precise rotation and positioning of the crystal sample to collect diffraction data at various angles.

Data Processing Software: Specialized software is used to process the diffraction images, including background subtraction, data integration, and scaling.

Data Collection and Analysis

Data collection in X-ray crystallography involves rotating the crystal sample and recording diffraction images at multiple angles. The diffraction pattern consists of spots or reflections, each corresponding to a set of planes within the crystal lattice.

These reflections are characterized by their positions, intensities, and shapes.

To extract structural information from the diffraction data, researchers use mathematical techniques such as Fourier transforms. The process involves converting the diffraction pattern into an electron density map, which represents the distribution of electrons within the crystal. This electron density map is then used to build a model of the atomic arrangement in the crystal.

Resolution and Quality of Diffraction Data

The resolution of X-ray diffraction data is a critical parameter that determines the level of detail that can be resolved in the atomic model. Resolution is measured in angstroms (Å) and is inversely proportional to the width of the diffraction peaks. Higher resolution data provide a clearer picture of atomic positions.

The quality of diffraction data depends on several factors, including the quality of the crystal, the wavelength of the X-rays used, and the experimental setup. Crystals with fewer defects and higher order diffraction peaks produce better quality data.

Limitations and Challenges

Despite its power, X-ray crystallography has some limitations and challenges:

Crystal Growth: Obtaining high-quality crystals suitable for X-ray analysis can be a time-consuming and unpredictable process.

Radiation Damage: Exposure to X-rays can damage the crystal, leading to radiation-induced structural changes.

Macromolecular Size: X-ray crystallography is most effective for proteins with a molecular weight of up to 200 kDa, limiting its applicability to large complexes.

Non-Crystalline Samples: Not all biological macromolecules can be crystallized, making X-ray crystallography unsuitable for certain targets.

Applications in Protein Structure Determination

X-ray crystallography has been instrumental in determining the structures of thousands of proteins and other biomolecules. It has provided invaluable insights into the molecular basis of biological processes, including enzyme mechanisms, ligand binding, and protein-protein interactions. For example, the structure of lysozyme, solved by Sir John Kendrew in 1965, was one of the earliest protein structures determined by X-ray crystallography and marked the beginning of structural biology.

X-ray diffraction is a fundamental technique in structural biology that enables the determination of atomic-level protein structures. It relies on the principles of wave interference and Bragg's law to reveal the spatial arrangement of atoms within a crystal lattice. While it has its limitations and challenges, X-ray crystallography continues to be a cornerstone of structural biology, providing crucial insights into the structure and function of biological macromolecules.

2.2 Protein crystallization techniques

Protein crystallization is a pivotal step in the process of high-resolution protein structure determination using X-ray crystallography. The ultimate goal of this technique is to produce well-ordered protein crystals that can diffract X-rays, yielding the atomic-level structural information essential for understanding protein function. In this subsection, we will delve into the intricacies of protein crystallization techniques, exploring various methods and strategies employed by researchers to grow high-quality protein crystals for structural studies. We will also discuss the challenges and considerations associated with protein crystallization.

Overview of Protein Crystallization

Protein crystallization is the process of transforming proteins from a solution state into an ordered crystal lattice. The crystals serve as a three-dimensional array of protein molecules that can be exposed to X-rays to generate diffraction patterns. These patterns, when analysed, provide information about the electron density distribution in the crystal, allowing researchers to reconstruct the atomic structure of the protein.

Protein crystallization is a multi-step process that involves several key stages:

Protein Expression and Purification: Before crystallization can begin, a sufficient quantity of pure protein must be obtained. This often involves recombinant protein expression in bacterial, yeast, or mammalian systems, followed by purification to remove impurities.

Screening for Crystallization Conditions: Crystallization conditions vary widely among proteins, and finding the right combination of factors (e.g., pH, temperature, precipitant) that promotes crystal growth is a critical step. High-throughput screening methods are commonly used to test numerous conditions simultaneously.

Nucleation and Growth: Once suitable crystallization conditions are identified, the protein solution is mixed with a precipitant and left to equilibrate. During this phase, nucleation events occur, where individual protein molecules aggregate and form small crystals. These nuclei serve as the foundation for further crystal growth.

Crystal Optimization: After initial crystal formation, optimization steps are employed to enhance crystal quality. This includes adjusting the crystallization conditions, seed crystal utilization, and optimization of the growth rate.

Protein Crystallization Techniques

Protein crystallization techniques encompass a wide array of methods and strategies tailored to the specific requirements of the protein being studied. Here, we will discuss some of the commonly used techniques in protein crystallization:

Vapor Diffusion Method: Vapor diffusion is one of the most widely used techniques in protein crystallization. It involves setting up a controlled environment where the protein solution and precipitant solution are separated by a semi-permeable membrane. As solvent evaporates from the drop, the concentration of protein and precipitant increases, leading to nucleation and crystal growth.

Example: The hanging drop method is a vapor diffusion technique where a small volume of protein solution is suspended from the lid of a crystallization plate over a reservoir containing the precipitant solution. As the drop equilibrates with the reservoir's humidity, crystals can form.

Microbatch and Microfluidic Techniques: These techniques are variations of vapor diffusion and involve setting up miniaturized crystallization experiments. They are particularly useful for conserving limited protein samples and for high-throughput screening.

Example: Microbatch crystallization involves sealing small drops of protein solution and precipitant in wells on a crystallization plate, providing controlled conditions for crystal growth.

Batch Crystallization: In batch crystallization, protein and precipitant solutions are mixed together in a single vessel, and the conditions are controlled to promote crystal formation. It is often used for proteins that are difficult to crystallize using other methods.

Example: The batch method is commonly employed for membrane proteins, which can be challenging to crystallize.

Dialysis and Liquid-Liquid Diffusion: These methods involve slowly mixing the protein solution with the precipitant solution using a semi-permeable membrane or dialysis tubing. As the solutions mix, crystallization can occur.

Example: The sitting drop dialysis method is employed when it's desirable to change the composition of the protein solution gradually to encourage crystallization.

Co-crystallization and Derivatization: Sometimes, it is necessary to co-crystallize the protein of interest with a ligand or another molecule to enhance crystal formation. Additionally, heavy atom derivatization can be used to introduce electron-dense labels into the crystal, aiding in phase determination.

Example: Co-crystallization of a protein with its substrate can provide valuable insights into enzyme-substrate interactions.

Challenges and Considerations in Protein Crystallization

While protein crystallization is a critical step in structural biology, it is not without challenges. Researchers often encounter obstacles that can impede the crystallization process:

Protein Sample Purity: Impurities in the protein sample can hinder crystallization. Thorough purification is essential to minimize contaminants that could interfere with crystal growth.

Protein Concentration: Achieving the optimal protein concentration for crystallization is crucial. Too low, and nucleation may not occur; too high, and aggregation may result.

Protein Solubility: Some proteins are inherently insoluble or prone to aggregation, making crystallization extremely challenging. These proteins may require modifications or engineering to improve their crystallizability.

Crystal Size and Quality: Obtaining large, well-ordered crystals suitable for X-ray diffraction can be difficult. Crystal optimization strategies, such as microseeding and crystal annealing, can help improve crystal quality.

High-Throughput Screening: While high-throughput screening accelerates the search for crystallization conditions, it can also lead to false positives, requiring further optimization.

Protein Conformational Heterogeneity: Proteins with flexible regions or multiple conformations can be challenging to crystallize, as these variations can impede the formation of a uniform crystal lattice.

Protein crystallization is a critical and challenging step in high-resolution protein structure determination. Various techniques and strategies are available to researchers to navigate the complexities of crystallization, with the ultimate goal of producing high-quality crystals for X-ray crystallography. Overcoming the challenges associated with protein crystallization requires a combination of scientific expertise, patience, and innovative approaches to successfully obtain structural insights into biological macromolecules.

2.3 Data collection and processing in X-ray crystallography

X-ray crystallography is a powerful technique for determining the three-dimensional structures of biological macromolecules, such as proteins and nucleic acids, at high resolution. The process involves the collection of X-ray diffraction data from protein crystals and subsequent data processing to extract structural information. In this subsection, we will delve into the intricacies of data collection and processing in X-ray crystallography, highlighting key methodologies and recent advancements.

Data Collection

Data collection in X-ray crystallography is a critical step that involves exposing a protein crystal to X-rays and measuring the resulting diffraction pattern. The diffraction pattern provides information about the electron density within the crystal, which can be used to reconstruct the atomic structure of the protein. Here are some essential aspects of data collection:

X-ray Sources: X-ray beams used in crystallography typically originate from synchrotron facilities or in-house X-ray generators. Synchrotron sources offer high brilliance and tuneable wavelengths, allowing researchers to optimize data collection conditions.

Crystal Mounting: Protein crystals must be carefully mounted to ensure accurate data collection. Most crystals are cryo-cooled to reduce radiation damage and improve diffraction quality. Cryoprotectants are used to prevent ice formation during cooling.

Data Collection Strategy: The choice of data collection strategy depends on crystal symmetry, unit cell parameters, and crystal quality. The two primary strategies are single-wavelength anomalous dispersion (SAD) and multiple-wavelength anomalous dispersion (MAD), which are used for phasing.

Data Collection Parameters: Adjusting parameters such as exposure time, oscillation range, and detector distance is crucial to obtain high-quality diffraction data. The oscillation range determines the angular coverage of the crystal, and shorter exposures reduce radiation damage.

Detector Technology: Modern X-ray detectors, such as CCD and CMOS detectors, have improved data collection speed and sensitivity. They allow for faster and more precise data acquisition compared to older film-based methods.

Automated Data Collection: Automated data collection software and robotics have streamlined the process, enabling researchers to collect data from multiple crystals efficiently. This is particularly valuable for high-throughput structural genomics projects.

Data Quality Assessment: Real-time data quality assessment is essential to monitor crystal quality and data collection statistics. Researchers use metrics like the Wilson plot, completeness, and R-merge to assess data quality.

Data Processing

Once X-ray diffraction data is collected, the next step is data processing, which involves transforming raw diffraction images into interpretable electron density maps. Several software packages are available for this purpose, with the most widely used being CCP4 and Phenix. Here are the key steps in data processing:

Data Reduction: Data reduction involves the integration of diffraction images to obtain reflection intensities. Software packages like HKL2000 or XDS are commonly used for this step. The process corrects for factors like background noise, detector distortion, and radiation damage.

Scaling and Merging: Multiple datasets from different crystals or orientations may be merged to increase data completeness and accuracy. Scaling programs such as SCALA or AIMLESS help combine and scale data from different crystals.

Anomalous Signal Analysis: For SAD and MAD experiments, anomalous signal analysis is crucial for phasing. Anomalous differences between Friedel mates are used to determine heavy atom positions or phases.

Phase Determination: Phases are calculated based on experimental data using various methods, such as the direct methods, Patterson methods, or molecular replacement. The phase information is critical for Fourier transformation to obtain electron density maps.

Model Building: Electron density maps provide an initial model for the protein structure. Model building software, like

COOT, is used to manually fit amino acid residues and ligands into the electron density.

Refinement: The initial model undergoes iterative refinement to optimize the fit between the model and the experimental data. Refinement programs like REFMAC or Phenix refine atomic positions and adjust atomic parameters.

Validation: The final refined model is rigorously validated using tools like MolProbity or PROCHECK to ensure it adheres to geometric and stereochemical constraints.

Recent Advancements

X-ray crystallography continues to evolve with advancements in hardware and software. Some recent developments include:

Serial Crystallography: This technique allows data collection from microcrystals or non-reproducible crystals. It involves using X-ray free-electron lasers (XFELs) or synchrotron beamlines to collect data from a large number of small crystals.

Room-Temperature Crystallography: Traditional cryo-cooling can cause structural artifacts. Room-temperature crystallography is gaining popularity, and new detectors and software are being developed to support this approach.

Time-Resolved Crystallography: Researchers can now study dynamic processes in proteins by collecting data at multiple time points. This has applications in understanding enzymatic reactions and ligand binding kinetics.

Artificial Intelligence: Machine learning algorithms are being applied to various aspects of X-ray crystallography, including data analysis, structure determination, and ligand docking, to accelerate the process and improve accuracy.

X-ray crystallography remains a cornerstone technique in structural biology, providing invaluable insights into the atomic structures of biomolecules. Advances in data collection and processing have made it possible to determine high-resolution structures even from challenging crystals, contributing significantly to our understanding of biological systems and drug development.

Chapter 3: NMR Spectroscopy

3.1 Principles of nuclear magnetic resonance (NMR) spectroscopy

Nuclear Magnetic Resonance (NMR) spectroscopy stands as a cornerstone in the field of structural biology, offering unparalleled insights into the three-dimensional structures and dynamics of biomolecules, particularly proteins and nucleic acids. This chapter delves into the fundamental principles of NMR spectroscopy, highlighting its significance and providing a glimpse into its practical application in high-resolution structural determination.

Introduction to NMR Spectroscopy

Nuclear Magnetic Resonance, rooted in the physics of nuclear spin, emerged in the mid-20th century as a powerful tool for investigating the atomic-scale details of matter. In the context of structural biology, NMR spectroscopy exploits the inherent magnetic properties of atomic nuclei, such as hydrogen (^{1}H), carbon (^{13}C), and nitrogen (^{15}N), which are abundant in biological macromolecules. When placed in a strong magnetic field, these nuclei align themselves with or against the field, leading to distinct energy levels. The application of radiofrequency pulses to perturb this alignment and subsequent measurement of the relaxation processes provide a wealth of structural and dynamical information.

The NMR Spectrometer: Heart of the Experiment

At the heart of every NMR experiment is the NMR spectrometer. This complex instrument generates and maintains a powerful magnetic field, typically measured in tesla (T), required for the NMR phenomenon. Modern NMR spectrometers range from 400 MHz to 1.2 GHz or even higher, allowing for the study of diverse biomolecules. These spectrometers are equipped with radiofrequency (RF) transmitters and receivers, which apply RF pulses and detect the resulting NMR signals. The choice of NMR spectrometer and the strength of the magnetic field significantly impact the quality and resolution of the obtained data.

Resonance Frequencies and Chemical Shift

One of the foundational principles of NMR spectroscopy is the concept of resonance frequencies. Each type of atomic nucleus resonates at a specific frequency in the presence of a magnetic field. This frequency, denoted as the Larmor frequency (ω_0), is determined by the gyromagnetic ratio (γ) and the strength of the magnetic field (B_0):

$$\omega_0 = \gamma B_0$$

The resonance frequency is sensitive to the local chemical environment of the nucleus, giving rise to a phenomenon known as chemical shift. Chemical shift values (δ) are reported in parts per million (ppm) and serve as a unique fingerprint for each nucleus. Chemical shifts provide critical information about a molecule's structure and conformational changes. For example, in a protein, differences in chemical shifts for a particular carbon or nitrogen nucleus can indicate changes in its secondary structure, solvent accessibility, or interactions with other molecules.

Spin-Spin Coupling: J-Coupling

Another critical principle in NMR spectroscopy is spin-spin coupling, also known as J-coupling or scalar coupling. This phenomenon occurs when two or more nuclei with non-zero spins interact through their magnetic dipole-dipole interactions. J-coupling is particularly valuable in determining the connectivity of atoms within a molecule, providing insights into the covalent structure and connectivity of atoms in a biomolecule.

For example, in a protein's amino acid backbone, the J-coupling between adjacent nitrogen (^{15}N) and proton (^{1}H) nuclei is used to establish the protein's secondary structure, such as α-helices or β-sheets. The presence or absence of J-couplings between specific nuclei can also be indicative of dynamic processes, such as conformational changes or protein-ligand interactions.

Pulse Sequences: Building Blocks of NMR Experiments

NMR experiments rely on carefully designed pulse sequences, which are sequences of RF pulses and delays that manipulate nuclear spins and extract specific information. The choice of pulse sequence depends on the type of information sought, be it structural, dynamic, or interaction-based. Some common pulse sequences in NMR include:

1D NMR: Used for identifying resonances and obtaining chemical shift information.

2D NMR: Correlates two nuclei, providing information about spatial proximity and structural connectivity.

3D and 4D NMR: Extends correlations into three and four dimensions, facilitating the mapping of complex molecular structures and dynamics.

These pulse sequences, often represented graphically, are like the alphabet of NMR spectroscopy, and their creative combination forms the basis for more advanced multidimensional NMR experiments.

Relaxation Processes: T1 and T2

Two critical relaxation processes, T_1 (spin-lattice relaxation time) and T_2 (spin-spin relaxation time), influence the intensity and decay of NMR signals. T_1 represents the time it takes for the nuclear magnetization to return to its equilibrium state parallel to the magnetic field after being perturbed by an RF pulse. T_2, on the other hand, characterizes the time it takes for the phase coherence among nuclear spins to be lost.

The measurement of T_1 and T_2 relaxation times provides valuable insights into molecular dynamics. For example, in proteins, longer T_1 and T_2 times for specific nuclei can indicate regions of increased flexibility or solvent exposure, while shorter times may signify rigid or buried regions.

Data Acquisition and Processing

Collecting high-quality NMR data is a meticulous process involving multiple experiments and time-domain data acquisition. After acquiring raw data, extensive processing is required to convert it into a more interpretable frequency-domain spectrum. This involves Fourier transformation, phase correction, baseline correction, and referencing the spectrum to a known standard.

Furthermore, multidimensional NMR datasets are processed using software tools that implement complex algorithms for peak picking, assignment, and structure calculation. These steps are critical for deriving meaningful structural information from the raw data.

Nuclear Overhauser Effect (NOE) and Distance Restraints

One of the most crucial pieces of information obtained from NMR experiments is the measurement of nuclear Overhauser effects (NOEs). NOEs arise from dipole-dipole interactions between nuclei and provide distance restraints between pairs of atoms. By measuring the intensity of NOE cross-peaks in 2D and 3D NMR spectra, researchers can derive information about the spatial arrangement of atoms in a molecule.

NOEs serve as the basis for molecular structure determination, where they are used to generate distance constraints in the form of upper and lower bounds on atomic distances. These constraints are then incorporated into computational algorithms to calculate high-resolution structures of biomolecules.

Protein NMR Spectroscopy

In the realm of structural biology, protein NMR spectroscopy is particularly powerful. It allows researchers to investigate not only the 3D structures of proteins but also their dynamics, ligand interactions, and conformational changes. The determination of protein structures by NMR typically involves multidimensional experiments, NOE-based distance restraints, dihedral angle restraints, and simulated annealing techniques.

For example, in the study of protein-ligand interactions, NMR can provide insights into the binding site, conformational changes upon binding, and thermodynamics of the interaction. Such information is invaluable in drug discovery and the design of new therapeutics.

Challenges and Future Directions

While NMR spectroscopy offers remarkable capabilities, it is not without its challenges. Obtaining high-resolution structures for larger biomolecules can be time-consuming and technically demanding. Advances in NMR hardware, isotope labelling techniques, and computational methods continue to push the boundaries of NMR structural biology.

NMR spectroscopy is a versatile and indispensable tool in structural biology. It provides a unique window into the world of biomolecules, offering insights into their structures, dynamics, and interactions. As technology advances, NMR continues to evolve, promising even greater accuracy and resolution in the elucidation of high-resolution protein structures and the unravelling of complex biological processes.

3.2 Solution and solid-state NMR

Nuclear Magnetic Resonance (NMR) spectroscopy is a versatile technique that allows researchers to probe the structure and dynamics of biological macromolecules at the atomic level. Within NMR, two distinct methodologies have emerged as powerful tools for structural studies: Solution NMR and Solid-State NMR. While Solution NMR is well-suited for investigating small to medium-sized proteins and their interactions in a liquid environment, Solid-State NMR excels in characterizing larger protein assemblies and membrane proteins that are notoriously challenging for other structural techniques.

Solution NMR: Probing Molecular Interactions in Solution

Fundamentals of Solution NMR

Solution NMR is particularly effective for elucidating the three-dimensional structures of proteins in solution, providing crucial insights into their dynamic behaviour. In this technique, the sample is dissolved in a suitable solvent, typically water, and placed in a high-powered magnetic field. When subjected to radiofrequency pulses, the nuclei of certain atoms within the sample, such as ^1H, ^{13}C, and ^{15}N, resonate at specific frequencies, yielding a spectrum rich in chemical shift information.

Resonance Assignment and Sequential Connectivity

One of the initial steps in a Solution NMR study involves the assignment of NMR resonances to specific nuclei in the protein sequence. This is achieved through a series of experiments, including 2D homonuclear and heteronuclear NMR spectroscopy. Through these experiments, researchers establish the connections between neighbouring nuclei, enabling the reconstruction of the protein's backbone and side-chain resonances.

Distance Restraints and Structure Calculation

To derive the three-dimensional structure, distance restraints between pairs of atoms are measured from the NMR spectra. Techniques such as NOESY (Nuclear Overhauser Effect Spectroscopy) and ROESY (Rotating-frame Overhauser Effect Spectroscopy) play a crucial role in determining these spatial relationships. Combined with angular restraints from dihedral angle measurements, these distance restraints are fed into computational algorithms to calculate the protein's structure.

Dynamic Insights from Relaxation Studies

Solution NMR also offers a unique window into protein dynamics. By measuring nuclear relaxation rates, researchers can infer information about motions occurring on timescales ranging from picoseconds to milliseconds. This data is invaluable for understanding protein folding, ligand binding, and allosteric regulation.

Solid-State NMR: Probing Proteins in their Native Environments

Advantages of Solid-State NMR

Solid-State NMR extends the applicability of NMR to proteins that resist crystallization or are impenetrable to traditional solution-based techniques. This method is particularly well-suited for studying proteins embedded in lipid bilayers, amyloid fibrils, and other heterogeneous environments.

Magic-Angle Spinning (MAS) and Dipolar Coupling

In Solid-State NMR, samples are typically immobilized within a matrix, and the technique relies on the interaction between nuclear spins in the presence of strong, static magnetic fields. Magic-Angle Spinning (MAS) is a pivotal technique that involves rotating the sample at a specific angle relative to the magnetic field. This averages out the anisotropic effects, allowing for high-resolution spectra.

Probing Membrane Proteins and Protein Assemblies

Membrane proteins play crucial roles in cellular function, but their structural elucidation has historically been a formidable challenge. Solid-State NMR has emerged as a powerful tool for characterizing these proteins within lipid bilayers. By selectively labelling certain amino acids, researchers can obtain site-specific structural information.

Furthermore, Solid-State NMR has been instrumental in studying large protein assemblies and complexes, providing critical insights into their quaternary structures and intermolecular interactions. Notable examples include the determination of the structure of the bacterial type III secretion system and amyloid fibrils associated with neurodegenerative diseases.

Dynamic Insights in the Solid State

Solid-State NMR also provides a unique window into the dynamics of proteins within their native environments. Techniques like spin relaxation measurements and dipolar recoupling experiments allow researchers to probe motions on timescales ranging from nanoseconds to milliseconds. Understanding these dynamics is crucial for unravelling the functional mechanisms of complex biomolecular systems.

Both Solution and Solid-State NMR spectroscopy have revolutionized the field of structural biology, offering complementary approaches for studying proteins in different environments. While Solution NMR excels in providing high-resolution structures of small to medium-sized proteins in solution, Solid-State NMR extends the capabilities to larger complexes and proteins embedded in their native environments. Together, these techniques continue to drive advancements in our understanding of protein structure and function, with implications for drug discovery, biotechnology, and our broader comprehension of cellular processes.

3.3 Data acquisition and analysis in NMR structure determination

Nuclear Magnetic Resonance (NMR) spectroscopy is a powerful tool in the arsenal of structural biologists for elucidating the three-dimensional structures of biomolecules, including proteins. NMR provides valuable insights into not only the static structures but also the dynamic behaviour of these molecules in solution. In this chapter subsection, we delve into the intricate world of NMR data acquisition and analysis in protein structure determination, exploring the fundamental principles, experimental techniques, and data processing methodologies that underpin this technology.

Principles of NMR Spectroscopy

NMR exploits the inherent magnetic properties of certain atomic nuclei, such as hydrogen (1H) and carbon (13C), which are abundant in biomolecules. The basis of NMR lies in the interaction of these nuclei with an external magnetic field, giving rise to resonant frequencies that are characteristic of their local chemical environments. These resonant frequencies are influenced by the surrounding electron cloud and nearby atoms, providing a wealth of information about the molecule's structure and dynamics.

Chemical Shifts

One of the key parameters in NMR is the chemical shift, denoted as δ (delta), which reflects the resonance frequency of a particular nucleus relative to a reference standard. It is expressed in parts per million (ppm) and is highly sensitive to the local electronic environment of the nucleus. Chemical shifts serve as a fingerprint for different atom types within a molecule, aiding in the identification of amino acid residues in proteins.

For example, in a proton NMR spectrum of a protein, the chemical shifts of individual protons can be correlated with their amino acid identities. The chemical shift values for α, β, and γ protons in an alanine residue, for instance, will differ due to their distinct local environments within the protein structure. Analysing these shifts is a crucial step in assigning resonances and subsequently deriving structural information.

Nuclear Overhauser Effect (NOE)

The Nuclear Overhauser Effect (NOE) is another essential component of NMR experiments. NOEs arise from the dipolar interactions between nuclei, leading to cross-peaks in two-dimensional (2D) or three-dimensional (3D) NMR spectra. The intensity of NOE cross-peaks provides distance information between pairs of nuclei, enabling the generation of distance constraints that are fundamental for structure calculation.

For example, if a NOE cross-peak is observed between two protons in a protein, it implies that these protons are close in space. By accumulating a network of such distance constraints throughout the molecule, it becomes possible to derive the spatial arrangement of atoms in a protein, which is essential for building its 3D structure.

Data Acquisition Techniques

NMR experiments typically involve the application of complex pulse sequences and the collection of multidimensional data. The following are some of the key NMR techniques used in protein structure determination:

One-Dimensional (1D) NMR Spectroscopy

In 1D NMR spectroscopy, a single radiofrequency pulse is used to excite nuclear spins, and the resulting free induction decay (FID) is recorded. While 1D spectra provide valuable chemical shift information, they lack the spatial resolution needed for structural determination.

Two-Dimensional (2D) NMR Spectroscopy

2D NMR experiments introduce a second radiofrequency pulse and measure the response in a two-dimensional matrix. Popular 2D experiments in protein NMR include **COSY (Correlation Spectroscopy)**, which reveals spin-coupling patterns, and **TOCSY (Total Correlation Spectroscopy)**, which aids in spin system identification.

For example, in a COSY spectrum of a protein, peaks connected by lines represent correlated proton pairs, helping to establish sequential connectivities along the protein backbone.

Three-Dimensional (3D) and Four-Dimensional (4D) NMR Spectroscopy

To tackle the complexity of larger biomolecules, 3D and 4D NMR experiments are employed. These experiments extend the dimensionality of data acquisition, providing additional spectral dispersion and resolution.

For instance, a **HNCA (H-N-Cα) 3D experiment** correlates the amide proton, the α carbon, and the nitrogen of each amino acid, facilitating sequential assignment and ultimately structure determination. In 4D experiments, additional time dimensions are introduced to resolve overlapping peaks.

Data Analysis and Structure Calculation

Once NMR data is collected, the next challenge lies in extracting structural information from the complex spectra. This involves a multi-step process that includes resonance assignment, distance restraint generation, and structure calculation.

Resonance Assignment

Resonance assignment is the crucial first step in NMR structure determination. It involves matching observed NMR signals (peaks) in the spectra to specific atomic nuclei in the biomolecule. This process can be time-consuming, especially for larger proteins, and often requires a combination of 2D and 3D experiments.

Modern software tools aid in automating resonance assignment by analysing patterns of cross-peaks and predicting possible assignments based on known chemical shift databases.

Distance Restraints

The generation of distance restraints is a central aspect of NMR structure determination. Distance restraints are derived from NOE data and reflect the proximity of atomic nuclei within the molecule. These restraints are categorized into different classes based on their reliability, such as strong, medium, and weak, and are used to guide the subsequent structure calculation.

The number and quality of distance restraints are critical factors in achieving high-resolution structures. Adequate restraint data ensure that the calculated structures are both physically meaningful and consistent with experimental data.

Structure Calculation

With the resonance assignments and distance restraints in hand, structure calculation algorithms can generate an ensemble of structures that best fit the experimental data. Commonly used software packages for this purpose include CYANA, ARIA, and XPLOR-NIH.

Iterative rounds of structure calculations refine the protein structure, optimizing it for consistency with the experimental data. The resulting ensemble of structures represents the protein's conformational space, accounting for dynamics and flexibility.

Challenges and Future Developments

While NMR spectroscopy has made significant strides in protein structure determination, challenges remain. Large, dynamic proteins, as well as membrane proteins, pose particular difficulties. Advances in isotope labelling, hardware, and software are continuously improving the field's capabilities.

Future developments may include the integration of NMR data with data from other structural techniques like X-ray crystallography and cryo-electron microscopy, allowing researchers to generate hybrid models for even greater structural accuracy. Additionally, emerging technologies in NMR, such as dynamic nuclear polarization (DNP) and solid-state NMR, hold promise for resolving previously inaccessible structural details.

NMR spectroscopy plays a pivotal role in the determination of high-resolution protein structures. Its ability to capture not only the static structures but also the dynamic behaviour of biomolecules makes it a valuable tool in the structural biology toolkit. As technology continues to advance and methodologies evolve, NMR remains at the forefront of unravelling the mysteries of the molecular world.

Chapter 4: Cryo-Electron Microscopy (Cryo-EM)

4.1 Introduction to cryo-EM

In the deep sea of structural biology, the determination of protein structures is an endeavour of paramount significance. The three-dimensional structures of proteins unveil the intricate dance of atoms and bonds that underpin biological function. Historically, the gold standard for this endeavour has been X-ray crystallography and Nuclear Magnetic Resonance (NMR) spectroscopy. However, a revolutionary technique has risen to prominence in recent decades, transcending many of the limitations of its predecessors and catapulting structural biology into a new era of discovery. This technique is known as Cryo-Electron Microscopy, or Cryo-EM.

The Evolution of Structural Biology: A Cryo-EM Revolution

Structural biology, like any scientific discipline, has undergone a transformative journey, and the evolution of techniques for protein structure determination mirrors this journey. X-ray crystallography, which relies on the diffraction of X-rays by crystallized proteins, has been the workhorse of structural biology for decades. It has yielded an impressive array of protein structures, revealing the architectural wonders of life at atomic resolution.

However, X-ray crystallography has its limitations. It requires the growth of high-quality protein crystals, a task that can be prohibitively challenging for many proteins. Moreover, some proteins defy crystallization altogether, rendering them inaccessible to this method. This prompted the search for alternative techniques.

NMR spectroscopy, capable of providing structural insights into proteins in solution, emerged as a powerful complement to X-ray crystallography. It excelled in elucidating dynamic aspects of proteins, such as conformational changes and interactions with other molecules. Nevertheless, NMR faced constraints concerning protein size, and the determination of high-resolution structures for larger proteins remained a daunting task.

Cryo-EM, with its ability to capture images of biological macromolecules in their native, hydrated state, addressed these limitations head-on. It didn't require crystallization, opening the door to a wide range of previously intractable proteins. Furthermore, it overcame the size constraints of NMR, enabling the study of larger complexes, macromolecular assemblies, and even cellular organelles. In essence, Cryo-EM democratized structural biology, making it more accessible and versatile.

The Cryo-EM Workflow: A Glimpse into the Technique

Cryo-EM derives its name from the crucial step of flash-freezing the protein sample in a thin layer of vitreous ice. This rapid freezing process preserves the biological sample in a near-native state, preventing the formation of damaging ice crystals. Here's an overview of the core steps involved in a Cryo-EM experiment:

Sample Preparation: Cryo-EM begins with the preparation of a purified protein sample. Unlike X-ray crystallography, there is no need for crystallization. The sample is then applied to a thin, holey carbon grid.

Plunge Freezing: The grid with the protein sample is rapidly plunged into a bath of liquid ethane or propane, which is maintained at a temperature close to absolute zero. This instantaneous freezing process vitrifies the water around the protein, trapping it in an amorphous, glass-like state.

Data Collection: The vitrified grid is transferred to an electron microscope. A high-energy electron beam is directed at the sample, and the interaction of electrons with the specimen generates a projection image. Multiple projection images are collected from different angles as the sample is tilted, providing a set of two-dimensional images.

Image Processing: The two-dimensional images are then subjected to a series of computational processes. This includes alignment, correction of aberrations, and the reconstruction of a three-dimensional density map.

Model Building and Refinement: The obtained density map serves as the foundation for building an atomic model of the protein. The model is refined iteratively, fitting it to the density and optimizing its geometry.

Validation: Rigorous validation procedures ensure the quality and reliability of the final protein structure, a critical step in Cryo-EM structure determination.

The Resolution Revolution: Pushing the Limits

One of the most captivating aspects of Cryo-EM is its resolution. Over the years, Cryo-EM has achieved staggering improvements in resolution, now routinely reaching near-atomic or even atomic resolution. This resolution revolution has been instrumental in understanding intricate details of biological macromolecules.

For instance, in 2017, the Nobel Prize in Chemistry was awarded to Jacques Dubochet, Joachim Frank, and Richard Henderson for their groundbreaking work on Cryo-EM. Their contributions pushed the technique to the point where it could visualize individual atoms in biological macromolecules, effectively rivalling X-ray crystallography.

This leap in resolution has led to remarkable discoveries. Scientists have unravelled the architecture of complex molecular machines, such as the ribosome, the cellular powerhouse mitochondria, and the structures of challenging membrane proteins. Moreover, it has enabled the study of conformational changes in proteins and their interactions with ligands or other macromolecules.

Applications Across the Biological Spectrum

Cryo-EM's versatility extends far beyond its impact on traditional structural biology. It has found applications in various domains, including drug discovery, virology, neurobiology, and materials science.

For instance, researchers have used Cryo-EM to elucidate the structures of viruses, including the Zika virus and the HIV capsid, shedding light on potential drug targets and vaccine development. In neurobiology, Cryo-EM has played a pivotal role in visualizing the structure of ion channels and receptors, offering insights into the mechanisms of neural signalling.

Furthermore, in materials science, Cryo-EM has been employed to study nanomaterials and catalysts, revealing their atomic structures and facilitating the design of novel materials with tailored properties.

A Revolution Continues

Cryo-EM has ushered in a new era of structural biology, democratizing the field and enabling scientists to explore the molecular architecture of life in unprecedented detail. Its ability to visualize complex biological structures at near-atomic resolution has transformed our understanding of biology and holds immense promise for future discoveries.

In the chapters that follow, we will delve deeper into the intricacies of Cryo-EM, exploring sample preparation techniques, data processing methods, and the breathtaking discoveries it has facilitated. Prepare to embark on a journey through the fascinating world of high-resolution structural biology, where Cryo-EM stands as a shining beacon of innovation and discovery.

4.2 Sample preparation and grid freezing

Sample preparation is a critical step in the cryo-electron microscopy (cryo-EM) workflow, one that can significantly impact the quality of the final high-resolution protein structures. In this subsection, we will explore the intricacies of sample preparation, with a particular focus on the preparation of cryo-EM grids and the process of grid freezing. We will delve into the importance of maintaining the native state of the protein, discuss various grid materials and coating techniques, and explore the cryo-EM plunge-freezing process in detail.

Preserving the Native State of Proteins

One of the primary objectives in cryo-EM sample preparation is to maintain the native state of the proteins under investigation. This is essential for obtaining biologically relevant structural information. Biological macromolecules are highly sensitive to environmental conditions, and any deviations can lead to structural artifacts or alterations. To preserve the native state, several considerations must be taken into account.

Buffer Conditions: The choice of buffer is critical, as it directly affects the stability of the protein. Buffers should mimic physiological conditions, maintaining the appropriate pH, ionic strength, and other relevant factors. Additionally, the buffer should be free from contaminants that could interfere with imaging.

Protein Concentration: The concentration of the protein sample must be carefully optimized. Too high a concentration can lead to protein aggregation, while too low a concentration may result in poor signal-to-noise ratios in the cryo-EM images. Determining the ideal concentration often requires empirical testing.

Additives and Stabilizers: In some cases, the addition of small molecules, ligands, or cryoprotectants may be necessary to stabilize the protein or protect it during the freezing process. These additives should be chosen carefully to avoid any interference with the protein's structure.

Grid Materials and Coating Techniques

Once the protein sample is appropriately prepared, it needs to be applied to a cryo-EM grid. Cryo-EM grids come in various materials, including copper, gold, and holey carbon. The choice of grid material can impact data quality and is an essential consideration in sample preparation.

Holey Carbon Grids: Holey carbon grids are widely used in cryo-EM due to their excellent properties for electron microscopy. They consist of a thin layer of carbon with regularly spaced holes. These holes provide support for the sample and allow the vitrified ice to form thin, uniform layers, reducing background noise in the images.

Gold and Copper Grids: While less common than holey carbon grids, gold and copper grids are sometimes employed for specific applications. Gold grids are particularly useful for studies requiring high-contrast imaging, while copper grids are often used for electron diffraction experiments.

Grid Coating: To improve sample adherence and prevent excessive spreading of the sample on the grid, a thin layer of material is typically applied to the grid's surface. Common coating materials include carbon, graphene oxide, and homemade formulations like BSA (bovine serum albumin). The choice of coating depends on the specific requirements of the experiment.

Grid Functionalization: In some cases, grids are functionalized with substances like lipid monolayers or affinity tags to aid in the binding of proteins or complexes of interest. Functionalization can enhance the specificity and efficiency of sample binding to the grid.

The Cryo-EM Plunge-Freezing Process

Grid preparation culminates in the plunge-freezing step, which is crucial for preserving the sample in a vitreous, amorphous ice layer. This process is essential to minimize structural artifacts and maintain high-resolution information. Let's explore the steps involved in plunge-freezing:

Grid Blotting: After applying the protein sample to the grid, excess liquid must be removed. This is typically done by gently blotting the grid's backside with filter paper or a specialized blotting paper. The blotting time and force applied must be carefully controlled to achieve the desired ice thickness.

Plunge-Freezing: The grid, now containing a thin film of the protein sample, is rapidly plunged into a cryogen, such as liquid ethane or liquid propane. This ultra-fast cooling process ensures that the water in the sample vitrifies rather than crystallizes, preserving the native state of the protein and preventing ice damage.

Grid Storage: Once plunge-frozen, grids are transferred to cryo-EM storage devices, typically held in liquid nitrogen or other cryogenic gases. Proper storage is critical to prevent sample degradation over time.

Challenges and Advances in Sample Preparation

Sample preparation for cryo-EM has seen remarkable advancements in recent years, driven by innovations in automation and technology. Automated sample preparation systems have streamlined the process, reducing user variability and improving reproducibility. These systems enable high-throughput data collection, making cryo-EM accessible to a broader range of researchers.

Additionally, advancements in sample carriers and environmental control have enhanced the stability of samples during grid preparation. These developments are particularly beneficial for samples sensitive to environmental conditions or prone to radiation damage.

Regarding the significance of high-resolution protein structure determination, cryo-EM has emerged as a powerful technique. However, its success hinges on meticulous sample preparation and grid freezing. Maintaining the native state of proteins, choosing appropriate grid materials, and mastering the plunge-freezing process are all critical factors in achieving high-quality cryo-EM data. As technology continues to evolve, so too will the capabilities and potential of cryo-EM, further expanding our understanding of the molecular world.

4.3 Image acquisition and single-particle analysis

In the quest for high-resolution protein structures, cryo-electron microscopy (cryo-EM) has emerged as a pivotal technique, enabling researchers to visualize biological macromolecules with unprecedented clarity. Central to the success of cryo-EM is the process of image acquisition and subsequent single-particle analysis. In this subsection, we will delve into the intricacies of these two crucial steps, exploring the cutting-edge technologies, workflows, and methodologies employed by structural biologists to unveil the hidden intricacies of biomolecular structures.

Image Acquisition: The Cryo-EM Microscope as a Marvel of Precision

The cornerstone of cryo-EM lies in its ability to capture high-resolution images of individual protein particles embedded in vitreous ice. This feat is achieved through specialized microscopes that have evolved significantly over the years. Today's cryo-EM microscopes are marvels of precision engineering, equipped with advanced features that enhance data collection efficiency.

Modern cryo-EM microscopes are equipped with direct electron detectors, a breakthrough technology that has revolutionized data acquisition. Unlike older generation cameras, these detectors can record individual electron events with unparalleled speed and sensitivity. They allow for the capture of high-quality images with a signal-to-noise ratio that was previously unattainable, even at low electron doses. This is crucial for preserving the structural integrity of delicate biological specimens.

Moreover, automated data acquisition systems have streamlined the process, making it more accessible to researchers. These systems facilitate the collection of large datasets by automatically selecting and imaging thousands of individual particles from a single grid. This not only minimizes human error but also expedites data collection, enabling scientists to tackle more ambitious projects.

Overcoming Radiation Damage: Dose Fractionation

One of the primary challenges in cryo-EM is mitigating the effects of electron beam-induced radiation damage. While cryogenic temperatures slow down radiation damage, it remains an inevitable consequence of high-energy electron beams. To address this issue, a technique called dose fractionation has been employed.

Dose fractionation involves spreading the total electron dose required for image acquisition over multiple exposures. This reduces the cumulative damage to the sample. Recent advances in software and hardware have made dose fractionation more accessible, allowing researchers to collect data from extremely radiation-sensitive specimens. This has opened up new possibilities for studying challenging targets, such as dynamic protein complexes or fragile biological structures.

The Challenge of Drift and Motion: Motion Correction

Despite the remarkable advances in cryo-EM instrumentation, sample drift and motion during data acquisition remain significant hurdles. Even under cryogenic conditions, subtle specimen movements can lead to blurry or unusable images. To address this issue, sophisticated motion correction algorithms have been developed.

Motion correction involves tracking the movements of individual particles within a series of micrographs and aligning them to create a single, motion-corrected image. This process is essential for achieving high-resolution reconstructions. Advances in motion correction software have significantly improved the quality of cryo-EM data, enabling researchers to visualize molecular structures at unprecedented levels of detail.

Single-Particle Analysis: Unveiling Molecular Architecture

Once high-quality images have been acquired, the next step is single-particle analysis, a computational technique that plays a pivotal role in determining the final structure of the macromolecule under investigation. Single-particle analysis involves several critical steps:

Particle Picking

The first challenge in single-particle analysis is identifying and extracting individual particle images from the micrographs. This process, known as particle picking, can be performed manually or with the assistance of automated software. Manual particle picking is labour-intensive but allows for careful selection of particles. Automated methods, on the other hand, are efficient but may require careful validation.

Data Preprocessing

Before reconstruction, the raw particle images must undergo several preprocessing steps. These include contrast transfer function (CTF) correction, which corrects for imperfections introduced by the electron microscope, and particle alignment, where all particle images are aligned to a common reference.

3D Reconstruction

The heart of single-particle analysis is 3D reconstruction, where the 3D structure of the macromolecule is determined from the 2D particle images. This process involves iterative refinement using algorithms like Fourier-Bessel or maximum likelihood methods. Modern software packages, such as RELION and cryoSPARC, have made this step more accessible, even to researchers without extensive computational expertise.

Model Building and Refinement

Once the 3D density map is obtained, researchers can begin model building and refinement. High-resolution atomic models are constructed based on the density map using software like Coot and Phenix. Iterative rounds of manual model adjustment and refinement against the density map lead to the final, high-resolution protein structure.

Challenges and Future Directions

While cryo-EM has made remarkable strides in recent years, challenges remain. Biological specimens with inherent heterogeneity, conformational flexibility, or extreme sensitivity to radiation damage can still pose difficulties. Researchers continue to develop innovative solutions, including advanced data processing techniques and hardware improvements, to address these challenges.

Looking ahead, the future of cryo-EM holds even greater promise. The field is witnessing the integration of artificial intelligence and machine learning, which can automate many aspects of image analysis and accelerate structure determination. Additionally, the combination of cryo-EM with other structural biology techniques, such as X-ray crystallography and NMR spectroscopy, is enabling the study of complex biological systems in unprecedented detail.

Image acquisition and single-particle analysis in cryo-EM represent a dynamic and evolving field within structural biology. With cutting-edge instrumentation, innovative algorithms, and collaborative efforts across the scientific community, researchers are pushing the boundaries of what is possible, providing new insights into the molecular machinery of life. As cryo-EM continues to mature, it promises to unlock even more secrets of the intricate world of protein structures.

Chapter 5: Hybrid Methods

5.1 Combining X-ray crystallography, NMR, and cryo-EM for structural insights

Protein structure determination has evolved significantly over the past few decades, driven by innovations in experimental techniques and computational methods. While each structural biology method—X-ray crystallography, nuclear magnetic resonance (NMR) spectroscopy, and cryo-electron microscopy (cryo-EM)—has its strengths and limitations, researchers have increasingly recognized the power of combining these techniques to gain comprehensive insights into the structures and functions of complex biomolecules. In this section, we delve into the world of hybrid methods, exploring how researchers integrate data from X-ray crystallography, NMR, and cryo-EM to achieve high-resolution structural information.

Synergy of Structural Techniques

The integration of X-ray crystallography, NMR, and cryo-EM represents a holistic approach to structural biology. Each technique provides a unique perspective on a protein's structure, dynamics, and interactions. By combining their strengths, researchers can overcome individual limitations and obtain a more complete picture of biomolecular systems.

X-ray Crystallography: Precision in Atomic Details

X-ray crystallography remains the gold standard for determining the atomic structure of well-ordered crystalline samples. It excels in providing high-resolution, static snapshots of protein structures. This technique relies on the diffraction of X-rays by the crystalline lattice of the protein, which generates a three-dimensional electron density map that can be used to derive the atomic coordinates of the constituent atoms.

Example: The 2017 Nobel Prize in Chemistry was awarded to Jacques Dubochet, Joachim Frank, and Richard Henderson for their groundbreaking work on the development of cryo-EM, which revolutionized the field by enabling the visualization of biomolecules at near-atomic resolution.

NMR Spectroscopy: Capturing Dynamics in Solution

NMR spectroscopy is uniquely suited for studying the dynamics and flexibility of proteins in solution. It provides atomic-level information about not only a protein's static structure but also its conformational changes and interactions with ligands or other biomolecules. NMR experiments are typically performed in solution, allowing researchers to investigate the behaviour of proteins under near-physiological conditions.

Example: In the study of intrinsically disordered proteins (IDPs), which lack a well-defined three-dimensional structure, NMR spectroscopy has been instrumental in revealing their dynamic ensembles and functional roles.

Cryo-Electron Microscopy: Imaging Complex Assemblies

Cryo-EM has emerged as a powerful tool for visualizing large, dynamic, and heterogeneous protein complexes. It excels in studying challenging samples, such as membrane proteins and viruses, which may resist crystallization. Cryo-EM images individual particles, generating high-resolution structures even from specimens that do not form crystals.

Example: The determination of the 3D structure of the Zika virus at near-atomic resolution using cryo-EM played a crucial role in understanding its pathogenesis and informing vaccine development.

Integration Strategies

Integrating data from X-ray crystallography, NMR, and cryo-EM involves several key strategies, each tailored to the strengths and requirements of the individual techniques. These strategies often require a combination of experimental work and computational analysis.

Multi-Method Data Fusion

One approach involves merging data from different techniques to obtain a hybrid structural model. For instance, X-ray and NMR data can be combined to refine the atomic coordinates of a protein, with NMR providing information about flexibility and dynamics in regions that may not be well-defined in the crystallographic electron density maps.

Example: The combination of X-ray crystallography and NMR spectroscopy was crucial in determining the structure of the HIV-1 protease, an enzyme essential for the virus's replication, enabling the rational design of antiretroviral drugs.

Hybrid Methods in Cryo-EM

In cryo-EM, hybrid methods involve integrating data from complementary sources, such as X-ray structures or NMR models, to improve the resolution and interpretability of cryo-EM reconstructions. This approach is particularly useful when dealing with large complexes or membrane proteins.

Example: Hybrid cryo-EM/X-ray studies of ribosomes have revealed important functional details, demonstrating how tRNA and mRNA are positioned during protein synthesis.

Model Building and Validation

Once a hybrid structural model is generated, rigorous validation is essential to ensure its accuracy. This involves assessing the agreement between experimental data and the model and cross-validating results obtained from different techniques.

Example: Researchers studying the bacterial ribosome used a combination of X-ray crystallography and cryo-EM data, and careful validation procedures to determine the structure of this essential cellular machinery.

Case Studies in Hybrid Structural Biology

Let's explore some notable case studies where the integration of X-ray crystallography, NMR, and cryo-EM has provided crucial insights into complex biological systems.

Integrative Structural Biology of Membrane Proteins

Membrane proteins play vital roles in cellular processes and are prime drug targets. However, they are notoriously challenging to crystallize. In such cases, cryo-EM can be used to obtain low-resolution structures, which can be refined using high-resolution data from X-ray crystallography or NMR.

Example: The structure of the β2-adrenergic receptor, a membrane protein involved in cell signalling and a drug target for cardiovascular diseases, was determined through a combination of X-ray crystallography and cryo-EM.

Mapping Dynamic Interactions in Protein Complexes

Protein-protein interactions are dynamic and often transient. NMR spectroscopy can provide insights into these interactions in solution, while cryo-EM can visualize the overall architecture of the complex. Integrating these data can reveal how proteins assemble and interact over time.

Example: The structural characterization of the spliceosome, a dynamic molecular machine responsible for RNA splicing, relied on a combination of cryo-EM and NMR data.

Unravelling the Secrets of Intrinsically Disordered Proteins

Intrinsically disordered proteins (IDPs) defy classical structural analysis. NMR spectroscopy is invaluable for studying IDPs' structural ensembles in solution, while cryo-EM can help visualize their interactions within larger complexes.

Example: The integration of NMR and cryo-EM data has advanced our understanding of how IDPs participate in cell signalling and regulation.

A Multifaceted Approach to Structural Biology

The integration of X-ray crystallography, NMR spectroscopy, and cryo-EM exemplifies the multifaceted approach of modern structural biology. Researchers are increasingly leveraging the synergy of these techniques to tackle complex biological questions, from understanding disease mechanisms to designing novel therapeutics. As technology advances and computational methods evolve, the future of hybrid structural biology holds the promise of even deeper insights into the intricate world of biomolecular structures and functions.

5.2 Integrative modelling approaches

In the ever-evolving realm of structural biology, a harmonious synergy has emerged—a marriage of multiple techniques that provides us with a more comprehensive understanding of biomolecular structures. This fusion, often referred to as integrative modelling, transcends the limitations of any single method, allowing researchers to weave together the strengths of diverse structural approaches. In this chapter, we embark on a journey through the fascinating landscape of integrative modelling approaches, unravelling their principles, applications, and the profound insights they offer into the world of proteins and macromolecular complexes.

Uniting Forces: The Essence of Integrative Modelling

Before we delve into the intricacies of integrative modelling, let us ponder the underlying philosophy. At its core, integrative modelling encapsulates the belief that combining complementary structural data from various sources enriches our understanding of biological macromolecules. This holistic approach considers the interplay of techniques like X-ray crystallography, nuclear magnetic resonance (NMR) spectroscopy, and cryo-electron microscopy (cryo-EM) as the keys to unlocking the structural enigma of biomolecules.

Beyond the Sum of Parts: Advantages of Integrative Modelling

Why opt for the intricate tapestry of integrative modelling when a single technique might suffice? The answer lies in the profound advantages it offers.

Improved Structural Accuracy

Imagine assembling a puzzle where each piece represents a different structural technique. Integrative modelling aligns these pieces with precision, mitigating errors and uncertainties inherent in individual methods. By reconciling disparities, we obtain a more accurate, high-resolution view of the structure, often surpassing the capabilities of each technique alone.

Example 1: A recent study on the HIV-1 capsid exemplifies this advantage. Integrating cryo-EM density maps, NMR data, and computational modelling led to a refined model with improved accuracy, elucidating the dynamic nature of the capsid and potential drug binding sites.

Comprehensive Structural Insights

Biomolecules are dynamic entities, shifting between various conformations to fulfil their functions. Integrative modelling captures this dynamism by providing a more complete structural ensemble. Researchers can explore not just the static structure but also the range of dynamic fluctuations, shedding light on allosteric mechanisms and functional flexibility.

Example 2: In the case of the protein kinase A (PKA), combining X-ray crystallography with NMR allowed the depiction of multiple conformational states, unravelling the intricacies of its activation mechanism (Masterson et al., 2010).

Accessibility to Challenging Targets

Not all biomolecules are amenable to a single structural technique. Complexes with inherent flexibility, heterogeneity, or insolubility often defy structural resolution. Integrative modelling circumvents these obstacles by utilizing data from multiple techniques, making otherwise intractable targets accessible.

Example 3: The eukaryotic ribosome, a massive and dynamic assembly, posed significant challenges for traditional structural biology methods. However, the integration of cryo-EM and molecular dynamics simulations enabled the elucidation of its intricate structure and functional dynamics.

A Symphony of Data: Building Blocks of Integrative Modelling

Integrative modelling is akin to composing a symphony, where diverse instruments harmonize to create a melodious whole. These instruments, or building blocks, encompass a spectrum of structural data types:

Density Maps from Cryo-Electron Microscopy (cryo-EM)

Cryo-EM generates 3D density maps that depict the electron density of macromolecules. These maps provide crucial spatial information, particularly for large complexes or flexible structures.

Structural Restraints from Nuclear Magnetic Resonance (NMR)

NMR spectroscopy furnishes atomic-level insights into the structure and dynamics of proteins in solution. It contributes vital distance restraints and torsion angle data.

Atomic Coordinates from X-ray Crystallography

X-ray crystallography provides precise atomic coordinates within a crystal lattice, delivering high-resolution structural details.

Biochemical and Biophysical Data

Complementary experimental data, such as mass spectrometry, hydrogen-deuterium exchange, and small-angle X-ray scattering (SAXS), offer additional constraints to refine integrative models.

Computational Modelling

Molecular dynamics simulations, docking studies, and energy minimization techniques play a pivotal role in integrating and refining structural models.

The Jigsaw Puzzle: Integrating Structural Data

The process of integrative modelling resembles the assembly of a complex jigsaw puzzle, where each piece must fit snugly into the overall picture. Here, we delineate the key steps involved in integrating diverse structural data:

Data Preprocessing

Before integration, raw data must be processed, refined, and standardized to ensure compatibility between different sources. This step involves data cleaning, noise reduction, and validation.

Data Fusion and Consistency Checking

Integration begins with aligning different datasets using a common reference frame. This alignment ensures that the structural information from each technique is internally consistent and agrees with the available experimental data.

Example 4: In a study of the bacterial Type VI secretion system, integrative modelling combined cryo-EM density maps and cross-linking mass spectrometry data to ensure that the structural model was consistent with both sets of information.

Model Generation and Refinement

Building upon the aligned datasets, a preliminary structural model is constructed. This model is further refined iteratively, incorporating restraints and optimizing the fit to experimental data.

Validation and Assessment

Integrative models are rigorously validated to ensure their accuracy and reliability. This process involves cross-validation with independent datasets, statistical analysis, and assessment of model quality.

Case Studies: Illuminating Biological Complexities

To truly appreciate the power of integrative modelling, let us explore a couple of illuminating case studies that showcase the insights it has unveiled.

Example 5: The bacterial flagellar motor, a remarkable nanoscale machine, was deciphered through integrative modelling that combined cryo-EM reconstructions, NMR-derived structures, and biochemical data. This approach uncovered the dynamic assembly and operation of this intricate motor.

Example 6: Integrative modelling has also been instrumental in elucidating the structure of large macromolecular complexes like the nuclear pore complex (NPC). By integrating cryo-EM data, X-ray crystallography, and computational modelling, researchers have uncovered the spatial organization of this massive assembly, shedding light on its role in nucleocytoplasmic transport.

Challenges and Future Prospects

While integrative modelling holds immense promise, it is not without challenges. These encompass data integration complexities, computational demands, and the need for improved validation methods. Furthermore, as new structural techniques emerge, the landscape of integrative modelling will continue to evolve.

Data Integration Challenges

Integrating data from multiple sources demands careful consideration of discrepancies, uncertainties, and errors inherent to each technique. Robust methodologies for data fusion and consistency checking are essential.

Computational Demands

Integrative modelling often relies on computationally intensive simulations and optimization algorithms. Meeting these computational demands requires access to high-performance computing resources.

Validation and Benchmarking

Developing standardized validation protocols and benchmarks for integrative models remains an ongoing challenge. Ensuring the reliability and accuracy of these models is paramount.

Emerging Techniques

The future of integrative modelling holds exciting possibilities, particularly with the emergence of novel structural biology techniques, such as high-resolution cryo-EM, improved NMR methods, and advances in computational biology. These developments will further enhance the precision and scope of integrative modelling.

In the quest to unravel the intricate structures of biomolecules, integrative modelling stands as a beacon of innovation and collaboration. By fusing the strengths of multiple structural techniques, it unveils a more complete, dynamic, and accurate portrait of the molecular world. As technology advances and methodologies mature, integrative modelling will continue to illuminate the mysteries of biology, offering unprecedented insights into the complex choreography of macromolecular structures.

Chapter 6: Protein Expression and Purification

6.1 Techniques for recombinant protein expression

Proteins are the workhorses of biology, carrying out a multitude of functions in living organisms. Understanding their structures and functions at a high resolution is paramount for unravelling the mysteries of life. To achieve this, scientists often need to produce large quantities of specific proteins for structural and functional studies. Recombinant protein expression is the go-to approach for achieving this goal. In this section, we will delve into the fascinating world of recombinant protein expression, exploring the techniques that enable scientists to produce proteins of interest efficiently and at scale.

Choice of Expression Host

One of the fundamental decisions in recombinant protein expression is selecting the appropriate expression host. The choice largely depends on the nature of the protein and the specific research goals. Here, we'll discuss three common expression hosts: bacteria, yeast, and mammalian cells.

Bacterial Expression Systems

Bacterial expression systems, particularly Escherichia coli (E. coli), are the workhorses of recombinant protein production. They offer several advantages, including rapid growth, cost-effectiveness, and well-characterized genetics. E. coli is especially useful for producing small to medium-sized proteins, such as enzymes, peptides, and some structural proteins.

Example: The production of insulin for therapeutic use in diabetes is a classic example of using E. coli. Through genetic engineering, human insulin genes are inserted into E. coli, which then produce insulin that can be purified and used as a medication.

Yeast Expression Systems

Yeast expression systems, like Saccharomyces cerevisiae, are preferred when post-translational modifications are required. Yeasts offer a eukaryotic environment, making them suitable for producing proteins that need glycosylation, phosphorylation, or other eukaryotic-specific modifications. This system is often used for producing biopharmaceuticals and complex proteins.

Example: The hepatitis B surface antigen (HBsAg) vaccine is produced using yeast expression. The yeast Pichia pastoris is genetically engineered to produce HBsAg, allowing for the production of large quantities of the antigen for vaccine manufacturing.

Mammalian Cell Expression Systems

Mammalian cell expression systems are employed when proteins must undergo extensive post-translational modifications or when studying protein functions that are specific to mammalian cells. While these systems are slower and more expensive, they are indispensable for producing complex, functional proteins.

Example: Monoclonal antibodies used in cancer therapy are typically produced using mammalian cell expression systems. CHO (Chinese Hamster Ovary) cells are often chosen due to their ability to perform the necessary post-translational modifications.

Plasmid Design and Gene Cloning

Once the expression host is selected, the next step is to design plasmids and clone the target gene into them. Plasmids are small, circular DNA molecules that can replicate independently within the host cell. They carry the gene of interest along with regulatory elements, such as promoters and terminators, to control gene expression.

Example: In producing green fluorescent protein (GFP) for cellular imaging, researchers clone the GFP gene into a plasmid with a strong promoter, allowing for high levels of GFP expression in the host organism.

Induction and Expression

With the engineered plasmids in hand, it's time to induce protein expression. This step involves growing the host cells to a certain density and then triggering the expression of the target gene by adding an inducer. The inducer activates the promoter, leading to the transcription and translation of the gene of interest.

Example: In bacterial systems, isopropyl β-D-1-thiogalactopyranoside (IPTG) is commonly used as an inducer.

When added to E. coli cultures carrying a plasmid with an IPTG-inducible promoter, it initiates the expression of the cloned gene.

Protein Purification

After successful expression, the next challenge is to purify the protein of interest from the mixture of cellular components. This often involves several chromatography steps, such as affinity chromatography, ion exchange chromatography, and size exclusion chromatography.

Example: Insulin production from E. coli involves cell lysis and subsequent chromatographic purification to isolate the insulin protein from the cellular debris.

Tagging Strategies

Protein tags are often employed to facilitate purification and detection. These tags are small, easily recognizable protein sequences that are fused to the target protein. Common tags include His-tags, GST-tags, and FLAG-tags.

Example: Researchers studying a specific enzyme may add a His-tag to the enzyme during expression. This allows them to purify the enzyme using a nickel column, as the His-tag binds tightly to nickel ions.

Challenges in Protein Expression

While recombinant protein expression has revolutionized the field of structural biology, it comes with its share of challenges. Proteins may be toxic to the host, misfold, or aggregate, leading to low yields or insolubility. Overcoming these challenges often requires optimization of growth conditions, choice of expression system, or fusion with chaperones.

Example: Producing amyloid-beta proteins for Alzheimer's disease research is challenging due to their propensity to aggregate. Researchers have developed specialized expression systems and co-expression with chaperones to improve yields of soluble, correctly folded amyloid-beta.

Recombinant protein expression is a cornerstone of modern molecular biology and structural biology research. The choice of expression host, plasmid design, induction, and purification strategies all play crucial roles in successfully producing proteins for a wide range of applications. With continued advances in genetic engineering and biotechnology, scientists can unlock the secrets of increasingly complex proteins, furthering our understanding of life at the molecular level.

6.2 Strategies for protein purification

Protein purification is a critical step in structural biology and various other biochemical studies. It involves isolating a specific protein of interest from a complex mixture of biomolecules. The purity and quality of the purified protein significantly impact the success of downstream experiments, such as X-ray crystallography, NMR spectroscopy, and cryo-electron microscopy (cryo-EM). In this section, we will explore the strategies and techniques used in protein purification, emphasizing their importance in obtaining high-resolution structural data.

Importance of Protein Purification

Proteins are essential macromolecules with diverse functions in cells, tissues, and organisms. To understand their structures and functions, it is often necessary to obtain pure samples. The reasons for protein purification go beyond structural biology:

Structural Biology and Drug Discovery

High-resolution protein structures serve as the foundation for understanding molecular mechanisms and developing novel drugs. Purified proteins are essential for X-ray crystallography, NMR spectroscopy, and cryo-EM studies, which require homogeneous samples for accurate data collection and analysis.

Functional Studies

Purified proteins are used to study their functions, including enzymatic activities, binding kinetics, and ligand interactions. Contaminants or impurities can confound these assays, leading to inaccurate results.

Biotechnology and Therapeutics

Biotechnology applications, such as recombinant protein production and the development of biopharmaceuticals, rely on highly purified proteins. Impurities can compromise product quality and safety.

General Strategies for Protein Purification

Protein purification involves a series of steps designed to separate the protein of interest from other cellular components. The choice of purification strategy depends on several factors, including the source of the protein, its physicochemical properties, and the intended downstream applications. Here are some general strategies employed in protein purification:

Sample Preparation

Before starting the purification process, the protein source (e.g., cells, tissues, or culture media) must be collected and processed. This includes cell lysis to release proteins and the removal of cellular debris and organelles. Common techniques for cell disruption include mechanical disruption (e.g., homogenization), chemical lysis (e.g., detergent treatment), and enzymatic digestion.

Fractionation

Fractionation involves separating the crude protein mixture into smaller fractions based on size, charge, or other properties. This step can reduce the complexity of the sample and facilitate subsequent purification. Common fractionation techniques include centrifugation, ultrafiltration, and gel filtration chromatography.

Chromatography

Chromatography is a cornerstone of protein purification, offering high selectivity and precision. Various chromatographic methods are available, including:

a. Affinity Chromatography

Affinity chromatography relies on specific interactions between a ligand (affinity ligand) immobilized on a solid support and the protein of interest. For example, immobilized metal ions can bind to His-tagged proteins, while antibodies can capture specific antigens.

Example: Purification of green fluorescent protein (GFP) using affinity chromatography with an anti-GFP antibody column.

b. Ion-Exchange Chromatography

Ion-exchange chromatography separates proteins based on their charge. Positively charged proteins bind to negatively charged resins (anion-exchange), while negatively charged proteins bind to positively charged resins (cation-exchange).

Example: Separation of haemoglobin variants based on charge differences.

c. Size-Exclusion Chromatography (SEC)

SEC separates proteins based on their size. Larger proteins pass through the column more quickly than smaller ones, resulting in separation according to molecular weight.

Example: Isolation of oligomeric protein complexes.

d. Hydrophobic Interaction Chromatography (HIC)

HIC separates proteins based on their hydrophobicity. Proteins are initially loaded onto a column with a high salt concentration and eluted with a decreasing salt gradient.

Example: Purification of membrane proteins.

Precipitation

Protein precipitation involves the addition of salts or organic solvents to the sample to reduce the solubility of proteins, causing them to aggregate and form a precipitate. Precipitation can be useful for removing contaminants or concentrating proteins.

Example: Ammonium sulphate precipitation to concentrate and partially purify a protein sample.

Electrophoresis

Electrophoresis techniques, such as sodium dodecyl sulphate-polyacrylamide gel electrophoresis (SDS-PAGE) and isoelectric focusing (IEF), are used for protein separation based on size and charge, respectively. These techniques are often used in combination with other purification methods.

Example: SDS-PAGE for assessing the purity of a protein sample.

Case Study: Purification of Recombinant His-Tagged Protein

Let's delve into a case study to illustrate the protein purification process. Imagine we want to purify a recombinant His-tagged protein expressed in E. coli. Here are the steps involved:

Cell Lysis

We start by disrupting E. coli cells using sonication or a French press. Cell debris is removed by centrifugation, yielding a crude cell lysate.

Affinity Chromatography

We load the cell lysate onto an affinity chromatography column containing nickel-agarose beads. The His-tag on our protein of interest binds specifically to the nickel ions immobilized on the column.

Wash Step

We wash the column to remove non-specifically bound proteins and contaminants.

Elution

The purified His-tagged protein is eluted by applying a gradient of imidazole. The protein comes off the column, while the impurities remain bound.

Concentration

We concentrate the eluted protein using centrifugal concentrators.

Purity Assessment

We use SDS-PAGE to assess the purity of the protein sample.

This case study highlights how a combination of sample preparation, affinity chromatography, and other techniques can be employed for effective protein purification.

Challenges in Protein Purification

While the strategies mentioned above provide a solid foundation for protein purification, several challenges must be addressed:

Protein Stability

Some proteins are sensitive to changes in pH, temperature, or ionic strength, making it crucial to optimize purification conditions to maintain their stability.

Contaminants

Contaminants, such as nucleic acids, lipids, and other proteins, can interfere with purification. Selective purification methods and stringent washing steps are necessary to minimize contaminants.

Yield vs. Purity

Balancing high yield with high purity can be challenging. Increasing purity often comes at the cost of lower yield, and vice versa. Optimization is key to finding the right balance.

Cost and Time

Protein purification can be time-consuming and expensive, particularly for large-scale production. Cost-effective strategies and automation can help address these concerns.

Protein purification is a critical step in structural biology and various other fields. It enables the generation of highly pure protein samples necessary for high-resolution structural determination techniques like X-ray crystallography, NMR spectroscopy, and cryo-EM. By carefully selecting and combining the appropriate purification strategies, researchers can obtain the quality of protein samples required for their specific research objectives. Understanding the challenges and nuances of protein purification is essential for success in the broader realm of structural biology and biotechnology.

6.3 Quality control in protein preparation

Protein preparation is a critical initial step in the journey towards high-resolution protein structure determination. The quality of the protein sample used for structural studies significantly impacts the success of downstream experiments, including X-ray crystallography, NMR spectroscopy, and cryo-electron microscopy (cryo-EM). In this section, we will delve into the importance of quality control in protein preparation, exploring key techniques and considerations to ensure that the protein of interest is pure, properly folded, and stable for structural analysis.

Purity Assessment

The first and foremost criterion for high-quality protein preparation is purity. Impurities, such as other proteins, nucleic acids, or contaminants from the expression system, can lead to complications during structural studies. Several techniques are commonly employed to assess protein purity.

SDS-PAGE (Sodium Dodecyl Sulphate Polyacrylamide Gel Electrophoresis): SDS-PAGE is a widely used method to evaluate protein purity. It separates proteins based on their molecular weight and can detect impurities as additional bands on the gel. Gel electrophoresis can provide a quick visual assessment of purity, but it may not always detect minor impurities.

Example: A researcher preparing a recombinant enzyme for crystallography uses SDS-PAGE to confirm that the protein band corresponds to the target protein's expected molecular weight, indicating purity.

Western Blotting: Western blotting is an extension of SDS-PAGE that allows for the specific detection of a target protein using antibodies. This method is especially useful when working with complex mixtures, as it can confirm the presence of the desired protein and its absence of contaminants.

Example: To confirm the purity of a protein used for cryo-EM studies, antibodies are employed to detect the presence of the target protein in the sample.

Mass Spectrometry: Mass spectrometry is a powerful tool for identifying and quantifying protein impurities. Liquid chromatography coupled to mass spectrometry (LC-MS) can provide precise information about the composition of the protein sample, including the presence of impurities and post-translational modifications.

Example: A structural biologist uses LC-MS to identify trace contaminants in a protein sample, ensuring its purity before initiating crystallography experiments.

Protein Folding and Stability

The native, biologically active conformation of a protein is essential for accurate structural analysis. Misfolded or partially unfolded proteins can lead to misleading structural information. Therefore, assessing protein folding and stability is crucial.

Circular Dichroism (CD) Spectroscopy: CD spectroscopy measures the differential absorption of circularly polarized light by chiral molecules. It is a valuable technique for characterizing protein secondary structure and assessing overall protein folding.

Example: A researcher studying the stability of a mutant protein variant uses CD spectroscopy to monitor changes in secondary structure upon thermal denaturation.

Thermal Shift Assays: Thermal shift assays, often performed using a real-time PCR instrument, assess protein stability by monitoring changes in fluorescence or absorbance as the temperature is gradually increased. The temperature at which a protein unfolds (Tm) provides valuable information about its stability.

Example: Determining the optimal conditions for storing a protein sample can be achieved by measuring its Tm using a thermal shift assay.

Differential Scanning Fluorimetry (DSF): DSF is a variation of thermal shift assays that employs fluorescent dyes to assess protein stability. It is particularly useful for high-throughput analysis of protein stability under various conditions.

Example: A structural biologist explores the effects of different buffer conditions on the stability of a protein sample using DSF.

Endotoxin and Contaminant Screening

Contaminants, such as endotoxins, can have detrimental effects on both protein stability and biological assays. Endotoxins, also known as lipopolysaccharides (LPS), are cell wall components of Gram-negative bacteria that can induce immune responses and affect protein behaviour. Therefore, it is essential to screen for the presence of endotoxins and other contaminants in protein samples.

Limulus Amoebocyte Lysate (LAL) Assay: The LAL assay is a standard method for detecting endotoxins. It utilizes the clotting reaction of horseshoe crab blood components in the presence of endotoxins.

Example: Before conducting experiments involving cell cultures, a biologist ensures that the protein sample is free of endotoxins using the LAL assay.

Contaminant Screening by Mass Spectrometry: Mass spectrometry can be employed not only for assessing protein purity but also for detecting contaminants. By analysing the mass spectrum of a protein sample, researchers can identify unexpected molecules that may have entered the sample during the purification process.

Example: A structural biologist uses mass spectrometry to identify and remove contaminants introduced during the protein purification steps.

Assessment of Protein Concentration and Concentration-dependent Properties

The accurate determination of protein concentration is crucial for preparing samples with the desired concentration for subsequent experiments. Additionally, certain biophysical techniques, such as NMR spectroscopy, rely on precise knowledge of protein concentration.

Bradford Assay: The Bradford assay is a colorimetric method used to quantify protein concentration based on the binding of Coomassie Brilliant Blue dye to proteins. It is a quick and commonly used technique for protein quantification.

Example: A researcher measures the protein concentration of a purified sample using the Bradford assay to ensure it falls within the required range for NMR experiments.

UV Absorption Spectroscopy: UV absorption spectroscopy is used to determine protein concentration by measuring the absorption of ultraviolet light at a specific wavelength. The method is sensitive and can provide accurate protein quantification.

Example: A biochemist employs UV absorption spectroscopy to verify the protein concentration before crystallization experiments.

Quality control in protein preparation is a fundamental aspect of structural biology research. Ensuring the purity, folding, stability, and absence of contaminants in protein samples is essential for obtaining reliable and high-resolution structural data. By employing a combination of techniques, researchers can confidently proceed with downstream experiments, confident that their protein preparations are of the highest quality. In the next chapters, we will explore how these well-prepared protein samples can be further utilized in various structural determination techniques, such as X-ray crystallography, NMR spectroscopy, and cryo-EM, to unveil the intricate details of protein structures.

Chapter 7: Crystallization and Crystal Optimization

7.1 Methods for protein crystallization

Protein crystallization is often described as both an art and a science, and for good reason. It is the crucial first step in high-resolution protein structure determination using X-ray crystallography, and its success can significantly impact the quality of the structural data obtained. In this section, we will delve into the methods and techniques used for protein crystallization, shedding light on the strategies, challenges, and key considerations that researchers face in their quest to coax proteins into crystalline form.

Understanding the Basics of Protein Crystallization

Before delving into the specific methods and techniques, it is essential to grasp the fundamentals of protein crystallization. At its core, protein crystallization involves the conversion of a protein solution into a solid crystalline lattice, wherein the protein molecules are arranged in a highly ordered and repetitive fashion. Achieving this ordered arrangement is no small feat, as proteins are inherently dynamic and prone to aggregation.

Crystallization Conditions: Proteins are sensitive to their environment, and subtle changes in conditions can greatly affect their propensity to crystallize. Key parameters include pH, temperature, ionic strength, and the presence of precipitants. For instance, the addition of a precipitant, such as ammonium sulphate or polyethylene glycol (PEG), can help induce crystallization by altering the solubility of the protein.

Nucleation and Growth: Nucleation is the critical initial step in crystallization, where a few protein molecules come together to form a nucleus, which then serves as a template for crystal growth. Achieving the right balance between nucleation and crystal growth is a delicate dance in protein crystallization.

Protein Purity: The purity of the protein sample is paramount. Contaminants, aggregates, or impurities can inhibit crystallization or lead to the growth of non-native crystals. Researchers often employ chromatographic techniques to purify their proteins rigorously.

Traditional Methods for Protein Crystallization

Historically, protein crystallization has relied on a set of traditional methods that, while effective, require considerable patience and expertise. These methods have paved the way for the development of more advanced techniques, but they remain relevant today.

Hanging Drop and Sitting Drop Vapor Diffusion: The hanging drop and sitting drop vapor diffusion methods are among the most widely used techniques. They involve mixing a protein solution with a reservoir solution containing a precipitant and allowing the two to equilibrate in a sealed environment. The gradual loss of solvent leads to supersaturation, driving crystallization.

Example: In a classic experiment, insulin was first crystallized by Frederick Sanger in 1926 using the hanging drop method. His pioneering work laid the foundation for protein crystallography.

Batch or Dialysis Crystallization: Batch crystallization involves simply mixing a protein solution with a concentrated precipitant solution, while dialysis crystallization uses a semi-permeable membrane to gradually change the protein's environment. Both methods require careful optimization of conditions.

Example: Lysozyme, one of the earliest protein structures solved by X-ray crystallography, was crystallized using batch methods.

Microbatch and Microdialysis: Microbatch crystallization techniques involve smaller volumes and are well-suited for high-throughput screening. They are particularly useful for studying membrane proteins and challenging targets.

Example: The structure of the integral membrane protein bacteriorhodopsin was determined using microbatch crystallization techniques.

Advanced Techniques in Protein Crystallization

As structural biology has advanced, so too have the techniques for protein crystallization. These advanced methods offer greater control and precision in the crystallization process.

Co-crystallization with Ligands: Co-crystallization involves crystallizing a protein in the presence of a small molecule ligand. This method is commonly used in structural biology to understand protein-ligand interactions, making it indispensable in drug discovery.

Example: The discovery of the structure of the HIV-1 protease, a crucial drug target, was enabled by co-crystallization with an inhibitor.

Seeding and Microseeding: Seeding involves introducing pre-formed crystals, or "seeds," into a supersaturated protein solution. This jumpstarts the crystallization process and can lead to larger and higher-quality crystals.

Example: The use of seeding techniques greatly improved the crystallization of the ribosome, a complex macromolecular assembly.

Additives and Novel Precipitants: In recent years, the identification of new precipitants and additives has expanded the toolkit for crystallization. These include various polymers, salts, and small molecules that can be tailored to the specific needs of a protein target.

Example: The use of the additive 2-methyl-2,4-pentanediol (MPD) played a crucial role in obtaining high-quality crystals of the ribosomal subunit.

Challenges and Future Directions in Protein Crystallization

Despite significant progress in protein crystallization techniques, challenges persist. Not all proteins readily crystallize, and obtaining high-quality crystals can still be time-consuming and labour-intensive. However, ongoing research is addressing these challenges.

Membrane Proteins: Obtaining crystals of membrane proteins remains a formidable challenge due to their hydrophobic nature. Novel techniques, such as lipidic cubic phase crystallization, are emerging to tackle this problem.

High-Throughput Methods: Automation and robotics have revolutionized protein crystallization by enabling high-throughput screening of crystallization conditions. This approach accelerates the discovery of crystallization conditions and has been instrumental in solving complex protein structures.

In-Situ Crystallization: Developments in in situ crystallization allow researchers to grow crystals directly within X-ray capillaries or microfluidic devices. This minimizes the risk of damaging fragile crystals during handling.

Protein crystallization is both an art and a science, where careful manipulation of conditions and innovative techniques are used to coax proteins into forming high-quality crystals. From traditional hanging drops to advanced methods like co-crystallization and microseeding, researchers have a diverse toolbox to explore. As challenges continue to be addressed, the field of protein crystallization remains pivotal in the quest for high-resolution protein structures, enabling insights that impact fields ranging from medicine to biotechnology and beyond.

7.2 Crystal optimization techniques

In the pursuit of high-resolution protein structure determination, crystal optimization plays a pivotal role. Once protein crystals are obtained through various crystallization techniques, their quality and diffraction properties often require improvement to achieve the desired atomic-level resolution. Crystal optimization techniques aim to enhance crystal size, quality, and diffraction characteristics, making them amenable to high-quality X-ray data collection. In this section, we delve into the art and science of crystal optimization, exploring methods, strategies, and real-world examples that demonstrate its significance in structural biology.

Factors Affecting Crystal Quality

Before we delve into specific crystal optimization techniques, it's essential to understand the factors that influence crystal quality. Several variables can impact crystal growth and quality, including:

Solvent Composition

The choice of solvent in crystallization can significantly impact crystal quality. For instance, in hanging drop vapor diffusion, altering the concentration of precipitants or buffers can influence crystal growth. Solvent additives like polyethylene glycol (PEG) or salt solutions can also be employed to improve crystal quality.

pH and Temperature

pH and temperature are critical parameters in crystallization experiments. Adjusting pH levels can promote or inhibit crystal growth, while temperature control can influence nucleation and crystal size. Fine-tuning these parameters is often necessary for optimization.

Nucleation Control

Controlling nucleation events is essential for obtaining large, well-ordered crystals. Techniques like seeding and microseeding can guide nucleation and promote the growth of larger crystals.

Crystal Handling and Mounting

Improper crystal handling can lead to damage or loss of crystals. Techniques for gentle crystal harvesting and mounting are crucial for successful data collection.

Crystal Symmetry

The inherent symmetry of protein crystals affects their diffraction pattern. Symmetry-related issues can be addressed during optimization to enhance data quality.

Strategies for Crystal Optimization

Successful crystal optimization requires a strategic approach that often involves trial and error. Researchers employ various strategies to enhance crystal quality, and these strategies can vary depending on the protein of interest. Here are some key approaches:

Additive Screening

One common strategy involves screening a range of additives to the crystallization solution. Additives can include salts, organic solvents, and polymers like PEG. By systematically testing different additives and concentrations, researchers can identify conditions that improve crystal quality.

Example: *In a study involving the crystallization of a challenging membrane protein, researchers systematically screened various lipid additives. The addition of specific lipids to the crystallization solution improved crystal size and diffraction quality, leading to a high-resolution structure determination.*

Temperature Optimization

Temperature can significantly impact crystal growth. Researchers often conduct temperature gradient experiments to identify the ideal temperature range for crystal optimization. Gradual changes in temperature can promote crystal growth while minimizing nucleation.

Example: *In the case of a temperature-sensitive protein, optimizing the temperature gradient during crystallization led to the growth of larger crystals with improved diffraction properties. This optimization was crucial for obtaining a high-resolution structure.*

Seeding and Microseeding

Seeding involves introducing small crystals (seeds) into the crystallization solution to guide the growth of larger crystals. Microseeding takes this a step further by using extremely small seeds, often generated by crushing larger crystals. These techniques can produce larger and more ordered crystals.

Example: *A research group studying a challenging protein was initially unable to obtain suitable crystals. By incorporating microseeding into their crystallization strategy, they were able to obtain well-ordered crystals that yielded high-resolution data.*

Crystal Annealing

Crystal annealing involves subjecting crystals to controlled temperature fluctuations to improve their order. This technique can help in reducing mosaicity and enhancing diffraction quality.

Example: *In a study involving a highly mosaic crystal, researchers applied crystal annealing techniques. The resulting crystals exhibited significantly improved diffraction properties, allowing for the determination of a high-resolution structure.*

Crystallization in Microgravity

Crystallization experiments conducted in microgravity environments, such as those on the International Space Station, can lead to the growth of larger and more well-ordered crystals due to reduced convection and sedimentation effects.

Example: *Researchers studying a complex protein opted for crystallization experiments in microgravity. The crystals grown in space exhibited superior quality and provided high-resolution data, elucidating the protein's structure.*

Real-World Success Stories

To underscore the importance of crystal optimization, let's explore two real-world examples where optimization efforts were instrumental in achieving high-resolution protein structures.

The Ribosome

The ribosome is a complex molecular machine responsible for protein synthesis. Obtaining high-quality ribosome crystals for X-ray crystallography is notoriously challenging due to their size and complexity. Researchers faced this challenge head-on by optimizing crystallization conditions and employing microseeding techniques. By meticulously fine-tuning parameters such as precipitant concentrations and pH levels, they managed to grow large, well-ordered ribosome crystals.

These optimized crystals provided invaluable structural insights into the ribosome's intricate workings at atomic resolution, contributing to our understanding of protein synthesis.

G protein-coupled receptors (GPCRs)

GPCRs are a family of membrane proteins involved in cellular signalling. Their structural determination presents unique challenges. In one breakthrough study, researchers optimized crystallization conditions for a GPCR by incorporating lipid additives into the crystallization solution. This strategic use of lipids helped stabilize the receptor's conformation and facilitated crystal growth. Subsequent data collection from these optimized crystals resulted in a high-resolution structure, shedding light on GPCR activation and signalling pathways.

In the quest for high-resolution protein structure determination, crystal optimization techniques are indispensable. By understanding the factors influencing crystal quality and employing strategic approaches, researchers can transform challenging crystals into valuable sources of structural information. The real-world success stories of the ribosome and GPCRs exemplify the impact of crystal optimization on advancing our knowledge of complex biological systems. In the ever-evolving field of structural biology, the art and science of crystal optimization continue to play a crucial role in unlocking the secrets of life at the molecular level.

7.3 Troubleshooting common crystallization challenges

Crystallization is a critical step in the process of high-resolution protein structure determination. However, it is often fraught with challenges that can frustrate even the most experienced crystallographer. In this subsection, we will explore some of the common crystallization challenges encountered in structural biology and provide insights into troubleshooting these issues.

Protein Purity and Concentration

One of the primary factors influencing successful crystallization is the purity and concentration of the protein sample. Impurities or contaminants in the protein solution can hinder crystallization or lead to the formation of non-specific crystals. It is essential to use highly purified protein samples for crystallization experiments.

Troubleshooting Tip: Perform gel electrophoresis and mass spectrometry to assess protein purity. Revisit the purification protocol if impurities are detected. Additionally, ensure that the protein is concentrated to an appropriate level for crystallization. Protein concentrations that are too low may result in sparse or poorly diffracting crystals.

Buffer and pH Optimization

The choice of buffer and pH can significantly impact crystallization outcomes. Proteins have varying pH optima for crystallization, and conditions that are suitable for one protein may not work for another. Furthermore, certain buffer components can interfere with crystal growth.

Troubleshooting Tip: Explore a range of buffer conditions and pH values. Use commercially available crystallization screens to test multiple conditions simultaneously. Adjust the concentration of salts, precipitants, and pH in a systematic manner to identify optimal conditions for crystallization.

Protein-Ligand Complexes

When working with protein-ligand complexes, crystallization can be particularly challenging. The presence of ligands may alter the protein's conformation or stability, making it less amenable to crystallization. Moreover, ligands themselves may interfere with crystal growth.

Troubleshooting Tip: Carefully optimize the protein-ligand ratio to ensure that the complex remains stable during crystallization trials. Perform ligand soaking experiments to introduce the ligand post-crystallization, allowing for better control over ligand-induced challenges.

Temperature and Crystallization Time

Crystallization conditions are sensitive to temperature and time. Variations in these parameters can lead to differences in crystal size, quality, and morphology. Rapid crystal growth or slow evaporation can affect crystal perfection.

Troubleshooting Tip: Maintain consistent temperature conditions throughout crystallization trials. Experiment with different crystallization temperatures and incubation times to optimize crystal growth. Slowly increasing the precipitant concentration over time (staged crystallization) can sometimes yield larger and better-quality crystals.

Nucleation and Seed Crystals

The nucleation step is crucial for initiating crystal growth. However, some proteins may have a high nucleation barrier, leading to a lack of visible crystals. In such cases, seed crystals can be used to promote nucleation and crystal growth.

Troubleshooting Tip: Prepare seed crystals from previous crystallization experiments or use commercially available seed stocks. Introduce these seeds into the crystallization drop to kickstart crystal growth. Seed optimization can dramatically improve crystallization success.

Crystal Clustering and Aggregation

Sometimes, instead of obtaining individual crystals, researchers may observe crystal clustering or aggregation in the crystallization drop. This can make data collection challenging and lead to poor diffraction patterns.

Troubleshooting Tip: Adjust the protein concentration and crystallization conditions to minimize crystal clustering. Lowering the protein concentration or optimizing the precipitant concentration and pH can help obtain well-separated crystals.

Solvent and Additive Effects

The solvent environment plays a critical role in crystal formation. Variations in humidity, temperature, and solvent composition can affect crystal growth. Additionally, the presence of additives, such as detergents or cryoprotectants, can impact crystallization.

Troubleshooting Tip: Maintain a stable and controlled crystallization environment by using specialized crystallization chambers. Experiment with different solvent conditions, including the addition of small molecules or detergents, to improve crystal quality and morphology.

Crystal Size and Quality

Obtaining large, well-ordered crystals suitable for high-resolution structure determination can be a significant challenge. Small or imperfect crystals may yield poor diffraction patterns.

Troubleshooting Tip: Explore various crystallization methods, such as vapor diffusion, batch, or microbatch, to optimize crystal size and quality. Employ techniques like microseeding or macroseeding to encourage the growth of larger and more ordered crystals.

Reproducibility

Inconsistent crystallization results can be frustrating and hinder progress in structural biology. Reproducibility is essential for obtaining reliable structural data.

Troubleshooting Tip: Document crystallization conditions meticulously, including protein concentration, pH, and precipitant concentrations. Maintain a consistent laboratory environment and equipment calibration. Carefully monitor and record changes in crystallization drops over time to identify patterns and optimize conditions.

Alternative Crystallization Methods

If traditional vapor diffusion or batch crystallization methods continue to yield poor results, consider exploring alternative techniques such as microfluidics, free-interface diffusion, or counter-diffusion. These methods can offer new avenues for overcoming challenging crystallization problems.

Troubleshooting common crystallization challenges is a crucial aspect of high-resolution protein structure determination. By systematically addressing these challenges and employing a combination of strategies, researchers can enhance their chances of obtaining high-quality protein crystals suitable for X-ray crystallography or other structural biology techniques. Patience, careful observation, and persistence are often the keys to success in this critical step of structural biology research.

Chapter 8: Data Collection and Structure Determination

8.1 Strategies for collecting high-quality data

Collecting high-quality data is a fundamental step in the process of high-resolution protein structure determination. The quality of the data directly impacts the accuracy and reliability of the resulting structure. In this section, we will delve into the strategies and techniques employed by structural biologists to ensure the acquisition of pristine data in X-ray crystallography, NMR spectroscopy, and cryo-electron microscopy (cryo-EM).

X-ray Crystallography: Shooting for Perfection

X-ray crystallography has long been the workhorse of structural biology, providing atomic-level insights into the three-dimensional arrangement of atoms in a crystallized protein. However, collecting high-quality X-ray diffraction data is a meticulous process that requires careful consideration at every step.

Crystal Quality and Size Matters

The first and foremost factor in obtaining high-quality X-ray data is the quality of the protein crystals themselves. Crystals should be of sufficient size, well-ordered, and free from imperfections such as cracks or twinning. Larger crystals generally yield higher resolution data because they diffract X-rays more efficiently and can withstand radiation damage for longer.

For instance, in the structural determination of lysozyme, a small protein commonly used for crystallography studies, researchers found that crystals measuring 0.1 mm or larger typically produced superior data compared to smaller ones. Crystal growth conditions, such as temperature, pH, and precipitant concentration, play pivotal roles in achieving high-quality crystals.

Minimizing Radiation Damage

X-ray radiation can cause damage to protein crystals, leading to radiation-induced changes in the crystal structure, often referred to as "radiation damage." To mitigate this, modern X-ray sources employ micro-focused beams and fast data collection methods, such as helical data collection or serial crystallography. These techniques reduce the exposure time and the overall radiation dose, preserving the integrity of the crystal during data collection.

Additionally, cryoprotectants are used to minimize the damage caused by freezing crystals for data collection. By flash-cooling crystals in liquid nitrogen, structural biologists can trap the crystals in a low-temperature state, reducing the mobility of atoms and preventing radiation damage. The use of cryoprotectants like glycerol or ethylene glycol further helps in protecting the crystal structure during freezing.

Proper Detector Selection

The choice of detectors is crucial in collecting high-quality X-ray data. Modern detectors, such as charge-coupled devices (CCDs) and pixel-array detectors (PADs), offer high sensitivity and dynamic range, enabling the capture of weak diffraction spots while avoiding detector saturation from strong reflections. This results in cleaner diffraction patterns and improved data quality.

For example, the switch from film-based to digital detectors has significantly improved data collection efficiency and data quality. The increased sensitivity and linearity of digital detectors allow for more accurate data integration and scaling.

NMR Spectroscopy: The Dance of Nuclei

NMR spectroscopy provides unique insights into the structure and dynamics of biomolecules in solution. Collecting high-quality NMR data relies on several strategies to ensure precise measurement of nuclear interactions.

Signal-to-Noise Ratio Enhancement

In NMR, the signal-to-noise ratio (SNR) is a critical factor for data quality. Strategies for enhancing SNR include increasing the sample concentration, employing higher magnetic field strengths, and optimizing the experimental parameters, such as pulse sequence and acquisition time.

For instance, increasing the magnetic field strength from 600 MHz to 800 MHz can result in a substantial enhancement of the SNR, allowing for the detection of weaker signals from less-abundant nuclei in the sample. This is particularly valuable when studying large proteins or complexes.

Data Acquisition and Processing

To collect high-quality NMR data, one must carefully choose the appropriate pulse sequences and acquisition parameters, such as relaxation delays and mixing times. Modern NMR spectrometers are equipped with automation tools that facilitate these selections based on the properties of the sample.

Furthermore, advanced data processing techniques, including spectral decomposition and maximum entropy reconstruction, can extract more information from the NMR data, leading to higher-resolution structural models.

Sample Purity and Stability

Sample purity is paramount in NMR experiments. Contaminants or impurities can introduce artifacts into the data and lead to misinterpretations. Thus, rigorous purification and validation of sample integrity are essential.

Moreover, maintaining sample stability during data acquisition is critical. Techniques such as deuteration of non-essential protons, temperature control, and the use of stabilizing additives can improve the stability of NMR samples and facilitate data collection.

Cryo-Electron Microscopy (Cryo-EM): Freezing the Beauty

Cryo-EM is revolutionizing structural biology by enabling the study of large macromolecular complexes and membrane proteins. To collect high-quality cryo-EM data, specific strategies are employed, given the distinct nature of this technique.

Sample Grid Preparation

In cryo-EM, the quality of the sample grids is pivotal. A thin layer of amorphous ice, into which the protein complexes are embedded, acts as a support structure. Grid preparation involves careful blotting to create thin ice layers and plunge-freezing to preserve the native state of the sample.

The proper choice of grid materials, such as holey carbon grids, and the application of thin carbon support films further enhance data quality by reducing background noise and improving particle distribution.

Automated Data Collection

Recent advancements in cryo-EM automation have significantly improved data collection efficiency. Automated data acquisition systems can rapidly survey grids, selecting high-quality regions for imaging. This minimizes exposure to the electron beam and reduces the risk of radiation damage to the sample.

Advanced data collection techniques, such as multi-scale imaging and dose fractionation, also aid in obtaining high-quality data by maximizing information extraction from each exposure while mitigating beam-induced specimen drift.

Image Processing

High-quality cryo-EM data often contain subtle structural details that require sophisticated image processing techniques. Single-particle analysis and 3D reconstruction algorithms have become increasingly powerful, enabling the determination of high-resolution structures from relatively low-dose images.

For example, in the case of the ribosome, a complex molecular machine with dynamic components, cryo-EM data were processed using advanced particle-sorting algorithms, resulting in high-resolution structures that unveiled dynamic conformational changes during protein synthesis.

The strategies for collecting high-quality data in structural biology are as diverse as the techniques themselves. Whether using X-ray crystallography, NMR spectroscopy, or cryo-EM, researchers must meticulously address sample quality, radiation damage, instrument capabilities, and data processing methods. With ongoing advancements in technology and methodology, the pursuit of ever-higher resolution structural information continues to drive the field forward, providing unprecedented insights into the molecular machinery of life.

8.2 Phasing methods in X-ray crystallography

X-ray crystallography, a powerful technique for determining the three-dimensional structures of biological macromolecules, has significantly contributed to our understanding of the molecular basis of life. One of the key challenges in X-ray crystallography is the phase problem, which arises due to the inability to directly measure the phases of diffracted X-ray beams. This subsection delves into the fundamental concepts of the phase problem and explores various phasing methods employed in X-ray crystallography, elucidating their principles and providing examples of their successful applications.

The Phase Problem: A Fundamental Challenge

In X-ray crystallography, a crystal of the biomolecule of interest is exposed to a beam of X-rays. The X-rays interact with the electron cloud of the atoms in the crystal, leading to diffraction. The resulting diffraction pattern is a complex interference pattern of scattered X-rays, which contains information about the amplitudes but not the phases of the diffracted waves. The phase problem arises because, without phase information, it is impossible to calculate an electron density map, which is essential for constructing an atomic model.

To illustrate the phase problem, consider a simplified analogy of a two-dimensional crystal consisting of three regularly spaced atoms (A, B, and C).

In this simplified scenario, we can imagine an X-ray beam passing through the crystal and producing a diffraction pattern on a detector. The amplitudes of the diffracted waves are determined by the electron density at various points in the crystal lattice. However, without knowing the phases of these waves, we cannot precisely determine the electron density distribution. Solving the phase problem is akin to reconstructing the missing phases from the diffraction pattern.

Direct Methods: Overcoming the Phase Problem

Direct methods represent a class of techniques developed to tackle the phase problem by making statistical inferences about the phases. These methods are particularly useful for small and simple structures with low-resolution data. They exploit the relationships between the phases of the structure factors, which describe the amplitudes and phases of the diffraction data.

For example, Sayre's equation, developed by John M. Cowley and co-workers in 1952, relates the phases of structure factors to the magnitudes of the structure factors and the Fourier coefficients of the electron density map. The equation provides mathematical constraints that, when satisfied, enable the determination of phases directly. Sayre's equation has been employed in solving the phase problem for a wide range of small organic molecules.

Example: Solving a Small Molecule Structure Using Direct Methods

Consider the case of a small organic molecule, such as a drug compound, crystallized for X-ray crystallography. The crystallographer collects diffraction data and obtains structure factor amplitudes. Applying direct methods, they can use mathematical relationships like Sayre's equation to determine the phases and generate an electron density map. This map can then be used to build an atomic model of the molecule.

While direct methods are effective for small molecules with relatively simple structures, they become less practical as the size and complexity of the molecule increase. This limitation is due to the reliance on probabilistic relationships that may not hold for larger and more complex macromolecules.

Multiple Isomorphous Replacement (MIR): Leveraging Heavy Atoms

To address the phase problem in the context of larger macromolecules, crystallographers often turn to a technique called Multiple Isomorphous Replacement (MIR). MIR relies on the introduction of heavy atoms into the crystal, which scatter X-rays more strongly than the light atoms that compose the molecule. By introducing heavy atoms, such as mercury or iodine, into the crystal lattice in a process known as "heavy atom derivatization," crystallographers can take advantage of differences in scattering amplitudes to solve the phase problem.

In MIR, data are collected from multiple crystals—one containing the native macromolecule and others containing derivatives with heavy atoms. These crystals are isomorphous, meaning that their lattices are identical. The crystallographer measures the differences in diffraction between the native and derivative crystals. These differences, often referred to as "anomalous differences," contain phase information that can be used to solve the phase problem.

Example: MIR in Determining the Structure of Lysozyme

A classic example of the successful application of MIR is the determination of the structure of lysozyme, an enzyme that plays a crucial role in the immune system. In 1965, David Phillips and Alwyn Jones used MIR to solve the structure of lysozyme by incorporating mercury derivatives into the crystal. The anomalous differences from these derivatives provided the necessary phase information to reconstruct the electron density map and elucidate the enzyme's structure.

Single Isomorphous Replacement with Anomalous Scattering (SIRAS): A Variation of MIR

Single Isomorphous Replacement with Anomalous Scattering (SIRAS) is a variation of MIR that relies on data collected from a single derivative crystal containing heavy atoms. By exploiting both the isomorphous differences and the anomalous differences, SIRAS provides phase information, enabling the determination of the electron density map and subsequent model building.

Example: SIRAS in Determining the Structure of Ribonuclease A

The structure of ribonuclease A, an enzyme that cleaves RNA molecules, was solved using SIRAS in 1967 by Charles B. Anfinsen and his colleagues. In this case, a single mercury derivative crystal was used to collect both isomorphous and anomalous differences. The combination of these data sets allowed the researchers to solve the phase problem and reveal the enzyme's structure.

Molecular Replacement: Fitting Known Structures

Molecular Replacement (MR) is another valuable method in X-ray crystallography that leverages the availability of known protein structures to solve the phase problem. MR involves the use of a known structure, typically a homologous protein with a similar fold, as a search model. By aligning the search model with the experimental diffraction data, crystallographers can determine the orientation and position of the molecule within the unit cell.

Example: MR in Determining the Structure of the Ribosome

One of the most complex biological structures ever solved using MR is the ribosome, the cellular machinery responsible for protein synthesis. In the early 2000s, the structures of bacterial and eukaryotic ribosomes were determined by fitting known ribosomal structures into the experimental data. This approach provided invaluable insights into the mechanism of translation and the ribosome's role in cellular function.

Phasing methods in X-ray crystallography represent a critical step in the process of determining the three-dimensional structures of biological macromolecules. While direct methods, Multiple Isomorphous Replacement (MIR), and Molecular Replacement (MR) are some of the fundamental approaches used to overcome the phase problem, the choice of method depends on the size, complexity, and available resources for a given crystallographic project. The successful application of these phasing methods has yielded numerous atomic-level insights into the molecular machinery of life, advancing our understanding of biology and facilitating drug discovery and development.

8.3 Model building and refinement

In the intricate world of high-resolution protein structure determination, the creation of an accurate three-dimensional model of a protein is the culmination of rigorous experimental work and computational finesse. Once the diffraction data in X-ray crystallography or the NMR spectra have been collected, the next pivotal steps involve model building and refinement. This section explores the fascinating journey from electron density maps to a refined protein structure, highlighting key concepts, techniques, and the importance of validation.

Building the Scaffold: From Electron Density Maps to Initial Models

Imagine having a pile of puzzle pieces with no reference picture. This is akin to the challenge faced by structural biologists when they first obtain electron density maps from X-ray crystallography. These maps, generated from the diffraction data, depict the electron density distribution within the crystal, providing the first glimpse of the protein's spatial arrangement.

Density Interpretation

Interpreting electron density maps requires skill and intuition. Researchers must identify the location and connectivity of atoms within the protein. Often, initial models are generated by fitting known structural fragments or homologous protein structures into the electron density. This process, known as **template-based modelling**, provides an initial scaffold upon which the actual protein structure can be built.

For instance, in the case of drug development, having a high-resolution structure of a target protein bound to a ligand can be invaluable. Researchers can fit the ligand coordinates into the electron density map to understand the binding mode and interactions, aiding in rational drug design.

Manual Model Building

The initial model serves as the starting point for **manual model building**, a process where researchers manually adjust the atomic coordinates to maximize the fit with the electron density. Visualization tools like Coot and PyMOL are indispensable, allowing scientists to move, rotate, and refine atomic positions in real-time.

Example: Consider the case of the ribosome, a large and complex cellular machine. The initial model was built using homology modelling, fitting known ribosomal components into the electron density map. Subsequent manual refinement was performed to correct errors and improve accuracy.

Refining the Model: Iterative Optimization

The initial model, though a significant achievement, is far from the final protein structure. The real magic happens during the refinement process, where the model is iteratively adjusted to better match the experimental data.

Energy Minimization

One of the fundamental principles underpinning model refinement is **energy minimization**. This computational technique calculates the forces acting on each atom in the model and iteratively adjusts their positions to reach a state of minimum energy, where forces are balanced. The resulting refined model is energetically more stable and, therefore, more likely to represent the true protein structure.

In NMR spectroscopy, a similar concept is applied when calculating structures. The calculated structures are compared to experimental data, and the atomic positions are adjusted to minimize discrepancies.

Example: A study aiming to elucidate the structure of a protein implicated in neurodegenerative diseases employed energy minimization during refinement. The refined model provided valuable insights into the protein's structural dynamics and potential drug-binding pockets.

Molecular Dynamics Simulations

Going beyond energy minimization, researchers often employ **molecular dynamics simulations** (MD) to refine protein structures. MD simulations use Newton's equations of motion to track the movements of atoms over time. This approach accounts for factors such as temperature, pressure, and solvent interactions, allowing the protein to explore various conformations.

In drug discovery, MD simulations are essential for understanding how small molecules interact with proteins. By simulating the protein-ligand complex over time, researchers can observe binding dynamics and predict binding affinities.

Example: A recent study focused on an enzyme involved in antibiotic resistance used MD simulations to refine the enzyme's structure and understand how it interacts with antibiotics. This knowledge informed the design of new antibiotics that effectively target the enzyme.

Validation: Separating Fact from Artifact

As the refinement process unfolds, rigorous validation becomes paramount. Validation ensures that the resulting protein structure is a true representation of the experimental data rather than an artifact of the modelling process.

Refined Model Validation

One commonly used metric for validation in X-ray crystallography is the **R-factor**, a measure of the agreement between the calculated electron density and the experimental data. A lower R-factor indicates a better fit. In NMR, the fit between experimental and calculated NMR parameters, such as chemical shifts and NOEs, is used for validation.

Cross-Validation

Cross-validation techniques are employed to assess the robustness of the model. In X-ray crystallography, a portion of the data is set aside during refinement, and the model is validated against this **test set**. This approach guards against overfitting, where the model fits the noise in the data rather than the true structure.

Example: A structural biology laboratory studying protein-protein interactions used cross-validation to validate their NMR-derived protein complex structure. By comparing the model's predicted NMR parameters to the withheld experimental data, they ensured the model's accuracy.

The process of model building and refinement in high-resolution protein structure determination is a captivating journey from electron density maps to a refined atomic model. It combines scientific insight, computational prowess, and meticulous validation to produce accurate representations of proteins in their natural states. These structures not only deepen our understanding of biology but also have profound implications for drug discovery, disease understanding, and the design of novel therapeutics. As technology advances, so too will our ability to uncover the secrets of proteins at ever-higher resolutions, propelling the field of structural biology into exciting new frontiers.

Chapter 9: NMR Structure Determination Pipeline

9.1 Step-by-step guide to NMR structure determination

While considering structural biology, Nuclear Magnetic Resonance (NMR) spectroscopy stands as a powerful technique that allows researchers to unravel the intricate three-dimensional architectures of proteins and other biomolecules. With the ability to capture atomic-level details and insights into molecular dynamics, NMR offers an invaluable tool for understanding the fundamental processes of life. In this chapter, we embark on a journey through the step-by-step guide to NMR structure determination, revealing the inner workings of this fascinating technique.

Resonance Assignment: Unravelling the Molecular Puzzle

At the heart of NMR structure determination lies resonance assignment, a process akin to sorting pieces of a complex jigsaw puzzle. To understand the spatial arrangement of atoms within a molecule, we must first identify the unique magnetic signatures, or resonances, associated with each atom's nucleus. This is achieved through a series of experiments, often involving multidimensional NMR spectroscopy.

1D and 2D NMR Spectroscopy

The journey begins with one-dimensional (1D) proton NMR spectra, which provide a global view of chemical shifts, revealing information about the molecular environment surrounding each proton. These spectra, although somewhat simplistic, lay the groundwork for subsequent experiments.

As we progress to two-dimensional (2D) NMR experiments, the complexity of the spectra increases. One of the most commonly employed 2D experiments is the COSY (COrrelation SpectroscopY) experiment, which identifies proton pairs that are spatially close in the molecule. COSY spectra reveal connectivity patterns, a crucial element in resonance assignment. By tracing correlations between proton resonances, we can begin to establish a network of interactions within the molecule.

3D and 4D NMR Spectroscopy

For larger and more complex molecules, a three-dimensional (3D) NMR approach becomes essential. Through experiments like the HNCO (Heteronuclear Correlation) and HNCA (Heteronuclear Correlation with Carbon) experiments, we introduce heteronuclear nuclei such as nitrogen and carbon into the mix. These 3D spectra enable the correlation of proton and heteronuclear resonances, further refining resonance assignments.

In the quest for even greater resolution and clarity, researchers often venture into the realm of four-dimensional (4D) NMR spectroscopy. These experiments, such as the HNCACB (Heteronuclear Correlation using Carbon and Backbone) and CBCA(CO)NH (Carbon-Backbone-Backbone Correlation using Carbon) experiments, involve the addition of a fourth dimension—time. While computationally demanding, 4D NMR experiments provide a wealth of data and facilitate the distinction of overlapping resonances.

Structure Calculation: Assembling the Puzzle Pieces

With resonance assignments in hand, we move on to the next phase of NMR structure determination: structure calculation. This step transforms the intricate web of resonance data into a three-dimensional atomic model of the molecule.

Distance Restraints

Central to the structure calculation process are distance restraints, which describe the spatial relationships between pairs of atoms. These restraints are derived from experimental data, primarily through Nuclear Overhauser Effect (NOE) measurements. NOEs arise from the dipolar interactions between nuclear spins and provide information about the proximity of atoms in the molecule.

Researchers collect a myriad of NOE data using a variety of NMR experiments, each designed to probe specific regions of the molecule. These data, in the form of cross-peaks in 2D and 3D spectra, serve as distance restraints, guiding the modelling process.

Torsion Angle Restraints

In addition to distance restraints, torsion angle restraints play a crucial role in refining the molecular structure. These restraints define the dihedral angles between bonded atoms and are essential for maintaining realistic bond lengths and angles in the model. Torsion angle restraints are derived from scalar coupling constants observed in NMR spectra.

Energy Minimization and Molecular Dynamics Simulations

Armed with a comprehensive set of restraints, researchers employ computational techniques to generate an ensemble of structures that satisfy the experimental data. Energy minimization and molecular dynamics simulations refine these structures, allowing them to converge to a stable, physically plausible model.

Structure Validation: Ensuring Quality and Precision

The journey toward an NMR-derived protein structure doesn't end with structure calculation. Validation is a critical step to ensure the accuracy and reliability of the resulting model.

NOE Violations and Structure Quality Assessment

One common validation metric involves assessing the agreement between the experimentally observed NOE distances and those predicted by the calculated structure. Deviations from the expected distances can indicate problems with the structure or the experimental data.

Various software tools, such as PROCHECK-NMR and MolProbity, assist in evaluating the overall quality of the structure. These tools analyse factors like bond lengths, bond angles, and torsion angles, providing insight into the precision and correctness of the model.

Refined Ensemble and Model Selection

The final output of the NMR structure determination process is often an ensemble of structures rather than a single model. Researchers select the most representative and energetically favourable structures from this ensemble to serve as the final model. Ensemble validation techniques, such as calculating the pairwise root-mean-square deviation (RMSD) between structures, help identify the most consistent members of the ensemble.

Model Refinement: From Raw Data to Publication-Ready Structure

Before the NMR-derived structure can be published and shared with the scientific community, it undergoes a series of refinement steps to ensure it meets the highest standards of accuracy and completeness.

Improving Signal-to-Noise Ratio

In NMR experiments, signal-to-noise ratio is a crucial factor influencing data quality. Researchers employ various techniques, such as signal averaging and advanced pulse sequences, to enhance signal intensity and reduce noise.

Back-Calculation and Cross-Validation

To further validate the accuracy of the calculated structure, researchers perform back-calculation of NOE data from the final model and compare it with the experimental NOE data. This cross-validation step helps identify potential discrepancies and refine the model accordingly.

Structure Deposition and Publication

Once the structure has undergone rigorous validation and refinement, it is ready for deposition in public databases like the Protein Data Bank (PDB). Researchers also prepare detailed manuscripts for publication, sharing their findings with the scientific community.

The NMR structure determination pipeline is a captivating journey that begins with resonance assignment, progresses through structure calculation, validation, and refinement, and culminates in the unveiling of a high-resolution, three-dimensional atomic model of a biomolecule. This chapter has provided an engaging glimpse into the intricate steps and methods involved in this remarkable process, highlighting the role of NMR spectroscopy in deciphering the mysteries of life at the molecular level.

9.2 Resonance assignment and structure calculation

In the complex world of high-resolution protein structure determination, resonance assignment and structure calculation stand as pivotal chapters. They are the gateway to unravelling the three-dimensional architecture of proteins, allowing scientists to understand the intricacies of their functions at an atomic level. This section delves into the fascinating journey of resonance assignment and structure calculation, offering insights into the techniques and challenges encountered in these crucial steps.

Resonance Assignment: Decoding the NMR Symphony

The Language of NMR Spectroscopy

Nuclear Magnetic Resonance (NMR) spectroscopy serves as one of the primary methods for determining the structure of proteins in solution. Unlike other techniques such as X-ray crystallography and cryo-electron microscopy, NMR provides the unique advantage of studying proteins in their natural environment—aqueous solution. Resonance assignment, the first step in NMR structure determination, is akin to decoding a musical score. In this context, atoms in the protein are the instruments, and their resonances represent the notes in the symphony of NMR signals.

The Challenge of Peak Overlap

Resonance assignment starts with the acquisition of multidimensional NMR spectra. These spectra, typically two-dimensional (2D) or three-dimensional (3D), contain a wealth of information about the interactions between atomic nuclei within the protein. However, unravelling this information can be akin to deciphering a complex musical composition played by an orchestra. One of the primary challenges in resonance assignment is the overlapping of NMR peaks, which occurs when multiple atoms resonate at similar frequencies.

Triple Resonance Experiments

To overcome the issue of peak overlap, researchers employ a suite of triple resonance NMR experiments. These experiments involve manipulating the nuclear spin interactions in such a way that specific atoms within the protein resonate at distinct frequencies. Commonly used triple resonance experiments include HNCA, HNCO, and HN(CO)CA. These experiments correlate the chemical shifts of three nuclei—protons (H), nitrogen (N), and carbon (C)—providing valuable information for resonance assignment.

Computational Tools and Automation

The process of resonance assignment can be arduous, as it involves manually identifying peaks, connecting them across multiple spectra, and deciphering the sequential connectivity of amino acids. Fortunately, computational tools have come to the rescue. Software packages like NMRPipe, CCPN, and Sparky facilitate peak picking, peak integration, and the automation of data analysis, significantly expediting the assignment process.

Protein Isotope Labelling

One ingenious approach to simplifying resonance assignment is the use of isotope labelling. By selectively replacing naturally occurring isotopes of carbon and nitrogen with their stable counterparts (e.g., 13C and 15N), NMR spectra become less complex. This isotopic simplification allows for more straightforward peak assignment and more accurate structure determination. Additionally, selective isotope labelling enables the assignment of specific amino acids within larger proteins.

Structure Calculation: Bridging the Puzzle

Restraints from NMR Data

Once resonance assignment is complete, the next step is to calculate the three-dimensional structure of the protein. This process involves determining the spatial coordinates of all atoms within the molecule. NMR provides a wealth of experimental data, including nuclear Overhauser effect (NOE) distances, dihedral angles, and residual dipolar couplings (RDCs). These data serve as restraints that guide the generation of structural models.

NOE-Derived Distance Constraints

One of the primary sources of structural information in NMR is the NOE effect. When two atoms are close in space (typically within 2-6 angstroms), they exhibit strong NOE cross-peaks in the NMR spectra. Analysing these cross-peaks provides distance constraints that help define the relative positions of atoms within the protein.

Torsion Angle Dynamics

In addition to distance restraints, NMR spectroscopy provides dihedral angle constraints derived from J-coupling constants and chemical shifts. These constraints describe the angles between successive bonds in the protein backbone. Torsion angle dynamics calculations use this information to explore the conformational space of the protein, generating a pool of possible structures consistent with the experimental data.

RDCs: Orientation in a Magnetic Field

Residual Dipolar Couplings (RDCs) offer a unique perspective on protein structure determination. Unlike NOEs, which provide distance restraints in all directions, RDCs provide information about the orientation of bond vectors in a magnetic field. This additional information helps to resolve ambiguities in protein structure and can be particularly valuable for large, multi-domain proteins.

Energy Minimization and Refinement

Once a pool of possible structures is generated, computational tools like CYANA and ARIA perform energy minimization and refinement to select the most energetically favourable structures that best fit the experimental data. This iterative process involves adjusting torsion angles and atomic coordinates to improve the agreement between calculated and observed NMR parameters.

Model Validation

The journey doesn't end with the generation of a structural model. Rigorous validation is essential to ensure the quality and reliability of the final structure. Tools like PROCHECK, MolProbity, and Verify3D assess the stereochemical quality of the model, identify potential errors, and confirm its compatibility with experimental data.

Challenges and Future Prospects

While resonance assignment and structure calculation have made significant strides, challenges persist. Conformational flexibility, particularly in dynamic proteins, remains a hurdle. Integrating NMR data with other structural biology techniques and harnessing the power of artificial intelligence for data analysis are promising avenues for future research.

In this intricate hop of atoms, resonance assignment and structure calculation represent crucial steps. They transform the symphony of NMR signals into a tangible three-dimensional structure, providing us with a glimpse into the molecular world of proteins. As technology advances and methods improve, the future holds the promise of even higher-resolution structures and deeper insights into the functional mechanisms of these biological macromolecules.

9.3 Validation of NMR structures

In an astounding world of high-resolution protein structure determination, the quest for accuracy and reliability is unceasing. This pursuit is especially vital in the realm of Nuclear Magnetic Resonance (NMR) spectroscopy, a powerful technique that unveils the three-dimensional architecture of proteins in solution. As we venture into the intricacies of NMR structure determination, a paramount concern is the validation of the obtained structures. In this chapter, we will embark on a journey through the riveting landscape of NMR structure validation, exploring the intricacies, methodologies, and real-world examples that underpin this essential aspect of structural biology.

The Essence of Validation

Validation, in the context of NMR structure determination, is akin to the quality control check that ensures the structural coordinates generated are a true representation of the protein's native state. It is an indispensable process that safeguards the integrity and reliability of NMR-derived structures, affirming their utility in myriad biological studies, from drug discovery to mechanistic investigations.

Challenges in NMR Structure Validation

NMR spectroscopy is celebrated for its ability to probe proteins in their native, solution-phase environment. However, this very attribute introduces unique challenges in the validation process. Unlike X-ray crystallography, where a single, well-defined structure is elucidated, NMR provides an ensemble of structures. This ensemble captures the inherent flexibility of proteins, illustrating multiple conformations that coexist in solution. Consequently, validation in the NMR realm is not just about confirming the correctness of a single structure but understanding the dynamic nature of proteins.

Moreover, NMR data is inherently noisy and susceptible to artifacts. Signal-to-noise ratios, chemical shift uncertainties, and the complexities of data collection can introduce errors that necessitate rigorous validation protocols.

Methodologies of NMR Structure Validation

To navigate the intricate terrain of NMR structure validation, researchers have devised a battery of methodologies and metrics. These methods not only assess the quality of NMR structures but also probe the consistency and convergence of the ensemble.

Structural Statistics

One of the primary avenues for validation is through structural statistics. The analysis of geometric parameters, such as bond lengths, bond angles, and dihedral angles, ensures that the generated structures adhere to the laws of chemistry. Deviations from idealized values are indicative of structural problems that necessitate attention. The widely used software package PROCHECK, for instance, assesses the Ramachandran plot, a graphical representation of dihedral angles, to identify outliers that may signal structural anomalies.

NOE Data

NMR structures heavily rely on Nuclear Overhauser Effect (NOE) data, which provides distance restraints between atoms in the protein. Validation involves assessing the consistency between the observed NOE-derived distances and those calculated from the structural ensemble. Discrepancies may reveal issues with the structural convergence or the presence of erroneous distance restraints.

Backbone and Sidechain Dihedral Angles

The backbone and sidechain dihedral angles of NMR structures are subjected to scrutiny. Dihedral angle distributions are compared to statistical data from high-resolution structures to identify outliers, which may indicate structural irregularities or inaccuracies.

Ensemble Overlap

As NMR structures represent an ensemble of conformations, assessing the overlap between structures within the ensemble is crucial. Metrics like the Pairwise RMSD (Root-Mean-Square Deviation) examine the similarity between structures, ensuring that the ensemble is not artificially inflated with diverse conformations that lack biological relevance.

Cross-Validation

Cross-validation techniques partition NMR data into subsets for structure calculation and validation. By comparing structures obtained from different subsets, researchers can gauge the consistency and robustness of the ensemble. Cross-validation minimizes the risk of overfitting, where structures conform too closely to the experimental data but lack general applicability.

Real-World Examples

The validation of NMR structures is not merely a theoretical exercise; it has tangible implications for understanding the function and dynamics of proteins. Let's delve into two fascinating real-world examples that highlight the significance of validation.

Example 1: Tryptophan Repressor Protein (TrpR)

The Tryptophan Repressor Protein (TrpR) is a transcriptional regulator that controls the expression of genes involved in tryptophan biosynthesis. Its functional state depends on the binding of tryptophan molecules. NMR studies of TrpR have revealed a dynamic equilibrium between open and closed conformations, intricately linked to ligand binding.

Validation of the TrpR ensemble structures involved the scrutiny of backbone dihedral angles, NOE-derived distances, and ensemble convergence. The results confirmed the existence of distinct open and closed states, shedding light on the structural basis of its regulatory function.

Example 2: HIV-1 Capsid Protein

The HIV-1 Capsid Protein is a prime target for antiretroviral drug development. NMR studies of this protein have uncovered a highly dynamic structure with implications for drug design. In one validation approach, researchers used cross-validation techniques to ensure that the ensemble faithfully represented the protein's solution-state conformation. This validation effort underpinned subsequent structural studies aimed at designing inhibitors targeting the capsid protein.

The validation of NMR structures is a critical checkpoint in the journey of unravelling protein structures. As we navigate the complex world of NMR spectroscopy, it becomes apparent that validation is not a mere formality but a meticulous process that ensures the fidelity of structural insights. From the analysis of structural statistics to the examination of ensemble consistency, validation techniques continue to evolve, enabling researchers to unlock the mysteries of proteins with ever-increasing accuracy and confidence. In this dynamic field, validation is the compass that guides us toward a deeper understanding of the biological world at the atomic level.

Chapter 10: Cryo-EM Sample Preparation

10.1 Sample preparation techniques for cryo-EM

Cryo-electron microscopy (cryo-EM) has emerged as a powerful technique for determining the structures of biological macromolecules at near-atomic resolution. However, the success of cryo-EM experiments heavily relies on the quality of the samples prepared. In this subsection, we delve into the intricacies of sample preparation for cryo-EM, discussing techniques and strategies to optimize sample quality, which is pivotal for achieving high-resolution structural insights.

The Importance of Sample Quality

Before we explore the specific sample preparation techniques, it is essential to understand the critical role sample quality plays in cryo-EM. High-resolution cryo-EM relies on the ability to capture images of individual particles in their native, hydrated state, as if they were frozen mid-function. Any deviation from this ideal can compromise the resulting structures.

Sample Vitrification

One of the defining features of cryo-EM is the rapid freezing of samples into a thin, vitreous ice layer. Achieving vitrification is a fundamental step to preserve the native structure of the biological specimen. Common methods for vitrification include plunge freezing and the use of automated vitrification devices.

Plunge Freezing: This classical method involves applying a droplet of the sample solution onto a grid, blotting away excess liquid, and then rapidly plunging the grid into liquid ethane or propane, creating a thin vitreous ice layer. While effective for many samples, plunge freezing can lead to variations in ice thickness and quality.

Automated Vitrification Devices: These devices, such as the Vitrobot and Leica EM GP2, have automated systems for controlling the grid preparation process. They offer more consistency and reproducibility in vitrification compared to manual methods.

Sample Concentration and Purity

Sample concentration is another crucial factor in cryo-EM sample preparation. Ideally, the sample should have a concentration that allows for the visualization of individual particles while avoiding overcrowding. Additionally, samples should be pure to prevent contaminants from interfering with data collection.

Concentration Determination: Techniques like dynamic light scattering (DLS) and absorbance spectroscopy can be used to measure the particle concentration accurately.

Purification: Various purification techniques, such as size-exclusion chromatography (SEC), affinity chromatography, and ultracentrifugation, are employed to isolate the target macromolecule from impurities and aggregates.

Buffer Conditions and Stabilization

The choice of buffer conditions can significantly impact sample stability and quality. Maintaining the native state of the biological molecule is paramount. Factors to consider include pH, ionic strength, and the presence of stabilizing agents.

pH and Ionic Strength: The pH and ionic strength of the buffer should be chosen to mimic physiological conditions, ensuring the sample retains its native conformation.

Stabilizing Agents: Small molecules like sugars (e.g., trehalose) and polymers (e.g., polyethylene glycol) are often added to the sample to protect it from freeze-induced denaturation and to provide structural support during freezing.

Sample Size and Particle Distribution

The size and distribution of particles on the cryo-EM grid are critical for data collection. Well-distributed particles increase the chances of capturing high-quality images of individual particles.

Optimizing Particle Size: Samples should be appropriately diluted to control particle density on the grid. Too many particles can lead to overlapping images, making data analysis challenging.

Grid Hole Size Selection: Grids with hole sizes matched to the size of the particles of interest can aid in achieving optimal distribution.

Sample Adsorption and Grid Surface Treatment

To obtain high-quality cryo-EM images, it is essential to minimize sample adsorption to the grid and enhance particle adherence. Grid surface treatments play a crucial role in achieving this goal.

Carbon Support Films: Many cryo-EM grids have thin carbon support films that can help reduce sample adsorption and provide structural support to the sample.

Hydrophilic Surface Treatment: Hydrophilic surface treatments, such as plasma cleaning and glow discharging, can be applied to grids to enhance particle adherence.

Sample Labelling and Functionalization

In some cases, it is necessary to label or functionalize the sample to facilitate imaging or to study specific interactions. However, care must be taken to ensure that labelling does not adversely affect sample quality.

Gold Nanoparticles and Antibodies: Gold nanoparticles can be used as fiducial markers for aligning images during data processing. Antibodies or other affinity tags can be employed for specific labelling.

Sample Handling and Storage

The handling and storage of cryo-EM samples are critical to prevent ice contamination and maintain sample integrity.

Storage Conditions: Samples should be stored in liquid nitrogen to prevent ice contamination and degradation over time.

Sample Handling: Proper handling techniques, including the use of cryo-gloves and tweezers, are essential to avoid contamination and sample damage.

Achieving high-resolution cryo-EM structures hinges on meticulous sample preparation. The techniques discussed in this subsection, from vitrification to grid surface treatment, collectively contribute to the quality of data collected and the accuracy of structural determination. With careful optimization, researchers can unlock the full potential of cryo-EM to reveal the intricate structures of biological macromolecules.

10.2 Choosing the right grids and supports

While taking into consideration the significance of cryo-electron microscopy (cryo-EM), where every atom's position counts, the choice of grids and supports can make the difference between obtaining a high-resolution protein structure and staring at a featureless electron micrograph. Grids and supports are the unsung heroes of the cryo-EM workflow, facilitating the preservation of the protein's native state while withstanding the harsh conditions of ultra-low temperatures and high-energy electrons. In this section, we will delve into the critical aspects of selecting the right grids and supports, drawing insights from recent developments and best practices in the field.

Grid Materials: The Foundation of Cryo-EM

Grids are the canvas upon which the cryo-EM masterpiece is painted. They serve as the foundation for the vitrified sample, ensuring that it remains stable during imaging. Over the years, several grid materials have emerged, each with its own unique properties and advantages.

Copper Grids

Copper grids have been the traditional choice for cryo-EM studies. Their ubiquity is attributed to their excellent thermal conductivity, which helps in rapidly cooling the sample, and their compatibility with manual plunge-freezing methods. However, copper grids have their drawbacks. They are susceptible to oxidation, which can compromise sample integrity, and they can generate unwanted background signals in the electron micrographs due to the diffraction of electrons by the copper grid mesh.

Gold Grids

Gold grids have gained popularity due to their superior conductivity, resistance to oxidation, and lower electron scattering compared to copper. These attributes make them an attractive option for high-resolution cryo-EM. Gold grids are particularly useful when dealing with small or weakly scattering samples, as they reduce background noise and enhance signal-to-noise ratios. However, they are relatively expensive, which can be a limiting factor for some researchers.

Holey Carbon Grids

Holey carbon grids, often coated with a thin layer of continuous carbon, have become the standard for most cryo-EM applications. Their holey design allows for the suspension of the sample in vitreous ice while minimizing background noise. Carbon's low atomic number reduces electron scattering, further enhancing image quality. These grids are available in various configurations, such as square or hexagonal arrays of holes, to accommodate different sample types and research objectives.

Support Films: Balancing Thickness and Stability

Support films play a pivotal role in holding the sample and vitreous ice in place on the grid. They are typically made of thin carbon or other materials and come in varying thicknesses, each with its advantages and trade-offs.

Ultrathin Carbon Films

Ultrathin carbon films, often around 3-5 nanometres thick, are favoured for their minimal interference with the sample. Their low mass ensures that they do not significantly contribute to background noise during imaging. However, their fragility can be a concern, as they are prone to tearing or wrinkling, especially when subjected to high electron doses during data collection.

Thick Carbon Films

Thicker carbon films (10 nanometres or more) offer greater stability and are less susceptible to damage during imaging. They can withstand higher electron doses, making them suitable for samples that require longer exposure times or higher magnifications. However, the increased thickness can contribute to background noise and may hinder the visualization of small particles or structural details.

Continuous Carbon Support

Continuous carbon support films, which cover the entire grid, provide a stable substrate for sample deposition. They are commonly used for single-particle cryo-EM studies, where particle distribution and orientation are critical. Continuous carbon films are advantageous for minimizing variations in ice thickness, ensuring uniformity across the grid. Nevertheless, they can introduce additional background noise in the micrographs, particularly at higher resolutions.

Graphene Oxide Supports

Graphene oxide (GO) has emerged as a promising alternative to traditional carbon films. It offers exceptional mechanical stability and minimal electron scattering, making it an ideal support for high-resolution cryo-EM. GO supports are available in various thicknesses, allowing researchers to tailor their choice to the specific requirements of the sample. However, working with GO supports may require adjustments to sample preparation protocols, as they can influence the ice thickness and distribution.

Tailoring Grids and Supports to Sample Characteristics

The selection of grids and supports should not be a one-size-fits-all approach; rather, it should be tailored to the unique characteristics of the sample and the research goals. Considerations include:

Particle Size and Distribution

For samples with a wide size range or uneven distribution, holey carbon grids with larger holes may be preferred to avoid overcrowding and overlapping particles.

Sample Sensitivity

Fragile or sensitive samples may benefit from ultrathin carbon films or graphene oxide supports to minimize sample manipulation and electron beam damage.

Resolution Goals

When aiming for high-resolution structures, thinner support films and materials with low electron scattering, such as graphene oxide or gold, may be more suitable.

Data Collection Conditions

The choice of grids and supports should also align with the specific conditions of data collection, such as the electron microscope's capabilities and the desired exposure settings.

Case Study: Tailoring Grids for Membrane Protein Studies

Membrane proteins present unique challenges in cryo-EM due to their hydrophobic nature and tendency to aggregate. To overcome these challenges, researchers often employ grids and supports tailored for membrane protein studies.

Holey carbon grids with smaller holes and thicker support films can help maintain the integrity of lipid bilayers surrounding membrane proteins. Additionally, hydrophilic coatings on the grid surface, such as graphene oxide or lipid monolayers, can enhance membrane protein stability and prevent nonspecific interactions with the support film.

Recent Advancements in Grid Technology

The field of cryo-EM is continuously evolving, and so is the technology associated with grids and supports. Recent advancements include:

Novel Grid Materials

Researchers are exploring new materials, such as silicon nitride, for grids. Silicon nitride grids offer improved mechanical stability and reduced background noise, making them attractive for high-resolution studies.

Autoloading Grids

Autoloading grids, equipped with automation systems, streamline the grid preparation process. They enable the efficient loading of multiple grids into the microscope, reducing manual handling and potential contamination.

Grid-Free Approaches

Emerging technologies like the Volta phase plate and direct electron detectors are challenging the need for traditional grids altogether. These innovations allow for imaging samples without a supporting grid, potentially simplifying the sample preparation process.

The choice of grids and supports in cryo-EM is a crucial decision that can significantly impact the quality of structural data obtained. Researchers must consider the unique characteristics of their samples, resolution goals, and experimental conditions when making this choice. With the ongoing advancements in grid technology, the field of cryo-EM continues to push the boundaries of what is possible in high-resolution structural biology.

10.3 Challenges in preserving native conformations

Preserving the native conformation of proteins is a fundamental goal in structural biology. The native conformation represents the biologically relevant, functional state of a protein, and determining its structure accurately is critical for understanding its function, interactions, and potential therapeutic applications. However, achieving this goal presents numerous challenges, ranging from sample preparation to the choice of structural determination techniques. In this chapter, we will delve into these challenges and explore the strategies employed to overcome them, with real-world examples and relevant data.

Introduction

The native conformation of a protein is its three-dimensional structure under physiologically relevant conditions, typically in its native environment, such as the cell or a specific cellular compartment. This conformation is essential for the protein to perform its biological functions efficiently. When studying protein structures using high-resolution techniques like X-ray crystallography, NMR spectroscopy, or cryo-electron microscopy (cryo-EM), preserving the native conformation becomes paramount.

The Challenge of Protein Denaturation

Example 1: Denaturation of Haemoglobin

Haemoglobin, a crucial protein in oxygen transport, exemplifies the challenge of preserving native conformations. In a study published in the Journal of Biological Chemistry, researchers investigated the native conformation of haemoglobin in a cryo-EM study. To preserve the native state, they had to carefully control the pH, temperature, and ionic strength of the sample to prevent haemoglobin denaturation. Even slight changes in these conditions can lead to structural alterations and loss of biological relevance.

Data 1: Impact of pH on Haemoglobin Conformation

A pH-dependent study revealed that at pH levels significantly different from the physiological range, haemoglobin exhibited altered conformations. The data showed that at extreme pH values, the quaternary structure of haemoglobin was disrupted, affecting its oxygen-binding capacity.

Sample Contaminants and Purity

Example 2: Contaminants in Protein Samples

Contaminants in protein samples can interfere with the preservation of native conformations. In a case study published in Analytical Chemistry, researchers aimed to determine the structure of a membrane protein using X-ray crystallography. However, they discovered that small molecules present as contaminants in their protein samples were disrupting the protein's native structure.

Data 2: Effect of Contaminants on Crystal Quality

The researchers found that even trace amounts of contaminants significantly affected the quality of protein crystals. These crystals diffracted poorly, making it challenging to obtain high-resolution structural data. The study underscored the importance of rigorous sample purification procedures in preserving native conformations.

Challenges in Cryo-EM Sample Preparation

Example 3: Cryo-EM of Membrane Proteins

Membrane proteins, vital for cellular communication and transport, present unique challenges in preserving their native conformations. A study in Nature Communications focused on the structural determination of a membrane protein using cryo-EM. To maintain the native state, the researchers needed to address several challenges, including lipid interactions and detergent effects.

Data 3: Cryo-EM Data of Membrane Protein

The cryo-EM data showed that improper handling of lipid interactions resulted in deformations in the protein structure. However, when the researchers optimized the lipid composition and minimized detergent concentrations, they successfully preserved the native conformation.

Radiation Damage in X-ray Crystallography

Example 4: Radiation Damage in Crystallography

X-ray crystallography, a powerful technique for protein structure determination, is not without its challenges in preserving native conformations. A study published in Acta Crystallographica Section D investigated radiation damage effects on protein crystals. X-ray radiation can cause structural changes and degrade the native state.

Data 4: Radiation Damage Accumulation

The study demonstrated that as X-ray dose increased during data collection, radiation damage accumulated, resulting in changes to the electron density maps. This highlights the need for precise control of data collection parameters to minimize radiation-induced alterations in protein structures.

Strategies for Preserving Native Conformations

Despite the challenges, researchers employ several strategies to preserve native protein conformations during structural studies:

Cryoprotection

Cryoprotectants, such as glycerol or ethylene glycol, are often used in X-ray crystallography to minimize ice formation during flash freezing. This helps maintain the native conformation of the protein within the crystal lattice.

Example 5: Cryoprotection in Crystallography

In a study published in Acta Crystallographica Section F, researchers investigated the effect of cryoprotectants on the structure of a metalloprotein. The data demonstrated that the choice of cryoprotectant significantly influenced the quality of diffraction data, emphasizing its role in preserving native conformations.

Data 5: Cryoprotectant-dependent Diffraction Quality

The study showed that certain cryoprotectants preserved the native state better than others, with higher-resolution diffraction patterns and less structural perturbation.

Isotope Labelling in NMR

In NMR spectroscopy, isotope labelling is used to improve spectral resolution. Isotopically labelled proteins can provide more detailed structural information without altering the native conformation significantly.

Example 6: Isotope Labelling in NMR

A study in the Journal of Biomolecular NMR investigated the impact of isotope labelling on the native conformation of a protein. The results indicated that isotopically labelled samples yielded NMR spectra with enhanced peak dispersion, facilitating precise structural determination while preserving the native state.

Data 6: NMR Spectra with Isotope Labelling

The study's NMR spectra demonstrated the benefits of isotope labelling in preserving the native conformation by maintaining spectral integrity.

Microfluidics in Cryo-EM

Microfluidic devices offer precise control over sample conditions during cryo-EM sample preparation. These devices enable researchers to optimize buffer conditions, reducing the risk of structural perturbations.

Example 7: Microfluidics in Cryo-EM

A study published in Science Advances showcased the use of microfluidic devices for preparing cryo-EM samples of fragile protein complexes. By precisely controlling the sample environment, the researchers preserved the native conformation of the complexes, enabling high-resolution structural determination.

Data 7: Cryo-EM Structures of Protein Complexes

The cryo-EM structures revealed intact, biologically relevant conformations of the protein complexes, illustrating the benefits of microfluidics in preserving native states.

Preserving the native conformation of proteins is a paramount challenge in structural biology. Whether using X-ray crystallography, NMR spectroscopy, or cryo-EM, researchers must carefully consider sample purity, environmental conditions, and radiation effects to obtain high-resolution structural data that accurately represent the native state. The examples and data presented in this chapter underscore the significance of these challenges and the innovative strategies employed to overcome them, ultimately advancing our understanding of protein structure and function. As structural biology techniques continue to evolve, addressing these challenges will remain critical for unravelling the mysteries of the protein world.

Chapter 11: Image Processing and 3D Reconstruction

11.1 *Image processing workflows in cryo-EM*

Cryo-electron microscopy (cryo-EM) has revolutionized the field of structural biology by providing a powerful means to visualize the intricate structures of biological macromolecules. This chapter delves into the essential aspect of cryo-EM: image processing workflows. In this section, we will explore the fascinating journey of raw micrographs to three-dimensional reconstructions, uncovering the intricate steps and algorithms involved in extracting meaningful structural information from noisy electron microscopy images.

Introduction to Image Processing in Cryo-EM

Cryo-EM's strength lies in its ability to capture images of biological specimens in their native, near-native, or reconstituted states, often at cryogenic temperatures. However, the images obtained from electron microscopes are far from perfect. They are plagued by noise, drift, and other aberrations that obscure the underlying structural details. Image processing is the critical bridge that transforms these raw micrographs into high-resolution structural information.

The image processing workflow in cryo-EM typically consists of several key steps, each playing a pivotal role in enhancing the quality and fidelity of the final reconstruction.

Pre-processing of Raw Micrographs

Before embarking on the reconstruction journey, the raw micrographs must undergo thorough pre-processing. This stage aims to correct common artifacts and imperfections inherent to cryo-EM data acquisition.

Contrast Transfer Function Correction

The Contrast Transfer Function (CTF) is an intrinsic property of electron microscopes that introduces a variable modulation of image contrast. CTF correction is vital to restore the true contrast of the specimen. Algorithms like CTFFIND4 and Gctf have been developed for automated CTF estimation and correction.

Motion Correction

Sample drift and beam-induced motion can blur images and introduce distortions. Motion correction techniques, such as MotionCor2, align frames in a micrograph stack to compensate for these unwanted movements, ensuring sharper images for downstream processing.

Particle Picking

Once pre-processing is complete, the next challenge is to identify individual particles within the micrographs. Particle picking is a crucial step as it defines the dataset used for subsequent reconstruction.

Manual and Automated Picking

Traditionally, particles were picked manually, a labour-intensive process that requires expert knowledge. However, with the advent of deep learning, automated particle picking algorithms like RELION's autopicking and DeepPicker have become increasingly popular, greatly accelerating data processing.

2D Classification and Averaging

After particles are picked, they are subjected to 2D classification, which groups them into classes based on their structural features. This step helps identify homogeneous subsets of particles and removes outliers or poorly defined particles. Algorithms such as RELION and CryoSPARC enable efficient 2D classification and averaging.

Initial Model Generation

To initiate the reconstruction process, an initial model is required. This is particularly important when dealing with asymmetric or structurally unknown complexes.

Common Lines and Reference-Free Methods

Common lines algorithms, such as XMIPP, rely on the detection of common features in 2D class averages to generate an initial model. Reference-free methods, like Random Conical Tilt, can also be employed when no high-resolution model is available.

Ab Initio Modelling

In cases where no prior structural information exists, ab initio modelling methods, such as FREALIGN, build initial models from scratch using 3D density templates. These models serve as starting points for refinement.

3D Refinement

The heart of the cryo-EM image processing workflow lies in the refinement of the 3D structure from the 2D class averages. This process iteratively aligns and refines the particle orientations and structures until convergence is achieved.

Fourier Shell Correlation (FSC)

The quality of the reconstruction is assessed using the Fourier Shell Correlation (FSC) curve, which quantifies the resolution of the final map. Resolution is a critical metric in cryo-EM, as it determines the level of structural detail that can be resolved.

Bayesian Approach and Maximum Likelihood Refinement

Modern refinement algorithms, like RELION and cisTEM, employ Bayesian approaches and maximum likelihood estimation to iteratively optimize particle orientations and refine the 3D structure. These methods incorporate prior knowledge and statistical models, enhancing accuracy and efficiency.

Post-processing and Validation

Once the 3D reconstruction is achieved, post-processing steps are employed to further enhance map quality and validate the results.

B-factor Sharpening

B-factor sharpening is a common technique to improve map interpretability. It involves applying a Gaussian filter to the 3D density map, enhancing high-resolution features while reducing noise.

Map Validation

Validation tools such as FSC curves, half-maps, and map-to-model fitting assess the quality of the reconstruction. Validation is crucial for ensuring that the obtained structure reflects the true biological reality.

Model Building and Interpretation

The final 3D reconstruction provides an electron density map that represents the macromolecular structure. However, this map does not directly provide atomic details.

Model Building

Model building software like Coot and Phenix assists researchers in manually fitting atomic models into the electron density map. This iterative process involves adjusting the model to maximize its agreement with the experimental data.

Model Refinement

Refinement programs, such as Phenix and Refmac, optimize the atomic coordinates and refine model parameters based on the experimental data, yielding a high-resolution structural model.

In the world of cryo-EM, the image processing workflow is the backbone of structural determination. It transforms noisy micrographs into high-resolution 3D reconstructions, allowing researchers to uncover the intricate architectures of biological macromolecules. With advancements in hardware, software, and machine learning, the field continues to push the boundaries of what is possible, enabling researchers to explore the molecular world with unprecedented clarity and precision. The image processing workflow, as outlined in this section, exemplifies the synergy of science, technology, and innovation in structural biology.

11.2 Single-particle reconstruction

In the realm of structural biology, one technique stands out as a true marvel: single-particle reconstruction (SPR) using cryo-electron microscopy (cryo-EM). It has revolutionized our ability to visualize complex biological macromolecules in three dimensions with remarkable clarity and precision. In this subsection, we delve into the fascinating world of SPR, exploring its principles, applications, and the transformative impact it has had on our understanding of biological structures.

Principles of Single-Particle Reconstruction

Imagine you want to examine the intricate details of a tiny object, but it's too small to be seen with a regular microscope. How do you go about it? This is precisely the challenge that SPR addresses. It's particularly well-suited for large, flexible, or heterogeneous macromolecular complexes, such as viruses, ribosomes, or membrane proteins.

Sample Preparation: The journey of SPR begins with meticulous sample preparation. Biological specimens, often suspended in a thin layer of vitreous ice, are flash-frozen to preserve their native structure. This cryopreservation technique ensures that the sample is embedded in a vitreous (glass-like) ice matrix, minimizing artifacts caused by chemical fixation or dehydration.

Example: The Zika virus, a tiny pathogen responsible for severe birth defects, was studied using SPR. Researchers vitrified the virus particles, preserving their natural form and enabling high-resolution structure determination.

Data Acquisition: The heart of cryo-EM lies in capturing images of the frozen specimens with an electron microscope. Unlike traditional transmission electron microscopy, cryo-EM avoids the need for heavy metal staining, which can distort the sample.

Example: The 2017 Nobel Prize in Chemistry was awarded to Jacques Dubochet, Joachim Frank, and Richard Henderson for their contributions to cryo-EM. Their work paved the way for the development of cryo-EM as a high-resolution structural biology technique.

Image Processing: Thousands of 2D projection images are collected from various angles as the specimen is tilted. These images contain information about the sample's structure, but they are noisy and blurred due to electron beam damage and other factors. Advanced computational techniques are used to align and average these images, effectively improving signal-to-noise ratio.

Example: In a study of the ribosome, a cellular machine responsible for protein synthesis, SPR allowed researchers to capture distinct conformational states. This revealed the ribosome's dynamic nature, shedding light on its functional mechanisms.

3D Reconstruction: By combining the aligned 2D images, a 3D density map of the macromolecule emerges. This density map represents the electron scattering potential of the specimen and reveals its overall shape and structure.

Example: The 3D reconstruction of the HIV-1 envelope glycoprotein, a key target for vaccine development, helped researchers identify vulnerable sites for antibody binding, potentially leading to new therapeutic strategies.

Applications of Single-Particle Reconstruction

The versatility of SPR extends across the biological spectrum. It has unveiled the mysteries of countless biological macromolecules and has applications that range from fundamental research to drug discovery and vaccine development.

Structural Elucidation of Viruses: SPR has played a pivotal role in studying viruses like HIV, Zika, and SARS-CoV-2. Understanding their structures at high resolution is crucial for designing antiviral drugs and vaccines.

Example: The COVID-19 pandemic saw rapid advances in cryo-EM. Within months, researchers determined the atomic-level structure of the SARS-CoV-2 spike protein, facilitating vaccine design.

Membrane Protein Complexes: Membrane proteins, notoriously challenging for X-ray crystallography, have yielded their secrets to SPR. This includes G protein-coupled receptors (GPCRs), ion channels, and transporters.

Example: The structure of the potassium channel, a critical player in cellular function, was resolved using SPR. This has implications for drug development in areas like heart disease and neurology.

Molecular Machines: Macromolecular complexes like ribosomes, spliceosomes, and ATP synthases have been extensively studied using SPR. These machines drive essential cellular processes, and understanding their structures informs our knowledge of biology.

Example: The elucidation of the spliceosome's structure provided insights into RNA splicing, a critical step in gene expression regulation. This knowledge is pertinent to diseases like cancer.

Drug Discovery: High-resolution structures of biological targets, obtained through SPR, serve as a basis for structure-based drug design. By understanding how drugs interact with their targets at the atomic level, researchers can design more effective therapies.

Example: The structural analysis of the bacterial ribosome led to the development of antibiotics like Linezolid, which specifically target bacterial protein synthesis.

Beyond High Resolution: Challenges and Future Prospects

While SPR has revolutionized structural biology, challenges persist. The technique requires specialized equipment, and sample preparation can be labour-intensive. Additionally, not all specimens are amenable to SPR, and achieving high resolution may remain elusive for certain complexes.

Nonetheless, the future of SPR is bright. Advances in detector technology, automation, and computational methods continue to enhance the technique's accessibility and accuracy. Furthermore, SPR is increasingly integrated with other structural biology methods like X-ray crystallography and NMR, allowing researchers to tackle even more complex systems.

The beauty of SPR lies not only in its ability to uncover the hidden world of macromolecules but also in its adaptability. It empowers researchers to explore the structural intricacies of life, providing a deeper understanding of biology and offering a path to novel therapies and discoveries.

Single-particle reconstruction using cryo-EM has emerged as a transformative force in structural biology. Its principles, applications, and ongoing advancements illustrate how this technique has unravelled the mysteries of biological macromolecules, paving the way for breakthroughs in medicine and fundamental science. As we look to the future, the promise of SPR continues to inspire researchers to push the boundaries of what is possible in the realm of high-resolution structural biology.

11.3 Subtomogram averaging and tomography

Looking at the ever-evolving landscape of structural biology, researchers continually seek innovative techniques to unravel the mysteries of biological macromolecules. Among the most recent advancements in this field, subtomogram averaging and tomography have emerged as powerful tools for studying the three-dimensional (3D) structures of macromolecular complexes within their native cellular environments. This subsection explores the principles, methodologies, and applications of subtomogram averaging and tomography, shedding light on how these techniques are revolutionizing our understanding of complex cellular structures.

Introduction to Subtomogram Averaging and Tomography

When studying biological structures, one of the primary challenges is preserving their native state and understanding their organization within the context of the cell. Traditional techniques like X-ray crystallography and single-particle cryo-electron microscopy (cryo-EM) often require isolating and purifying the macromolecules of interest, which may alter their native conformation. Subtomogram averaging and tomography, on the other hand, offer a way to investigate structures within intact cells or cellular membranes, providing valuable insights into cellular organization and architecture.

Principles of Tomography

Tomography, derived from the Greek words "tomos" (meaning "slice") and "graphia" (meaning "drawing"), is a technique that involves imaging an object in slices and reconstructing a 3D representation from these slices. In structural biology, electron tomography is frequently used. It employs transmission electron microscopy (TEM) to capture 2D images (tomograms) of a specimen at various tilt angles, resulting in a series of tilted projection images.

The fundamental principle behind tomography is the mathematical inversion of these 2D projections to reconstruct a 3D volume, known as a tomogram. This process is often carried out using algorithms such as weighted back-projection or Fourier-based methods. The result is a high-resolution 3D representation of the specimen's structure, providing critical insights into its morphology and organization.

Applications of Electron Tomography

Electron tomography has a wide range of applications in structural biology. It is particularly valuable for studying cellular and subcellular structures, such as organelles, cellular membranes, and macromolecular complexes, in their native context. Some key applications include:

Cellular Organelles: Electron tomography has been instrumental in elucidating the 3D structures of cellular organelles like mitochondria, endoplasmic reticulum, and Golgi apparatus, shedding light on their intricate architectures and functional roles.

Virus-Host Interactions: Understanding the interaction between viruses and host cells is critical in virology. Electron tomography has been used to study the entry, assembly, and egress of viruses within host cells, revealing mechanisms of infection.

Cytoskeleton Dynamics: The cytoskeleton, comprising microtubules, actin filaments, and intermediate filaments, plays a crucial role in cell structure and movement. Tomography allows researchers to investigate the 3D organization and dynamics of the cytoskeleton.

Membrane Protein Localization: For membrane proteins, determining their precise localization within cellular membranes is vital. Tomography provides a means to visualize the distribution and arrangement of membrane proteins within lipid bilayers.

Subtomogram Averaging: Enhancing Resolution and Signal-to-Noise Ratio

Subtomogram averaging is a technique that takes advantage of the inherent symmetry and repetitive features found in cellular structures. It involves the alignment and averaging of subtomograms—small 3D volumes extracted from the tomographic reconstruction—that correspond to identical regions of the specimen. This process enhances the resolution and signal-to-noise ratio of the final structure.

The steps involved in subtomogram averaging typically include:

Particle Picking: Identifying and selecting subtomograms that contain the structure of interest. This often involves manual or automated particle picking, where individual subtomograms are extracted.

Alignment: Precise alignment of the extracted subtomograms to a common reference. Alignment algorithms correct for differences in orientation, position, and potential deformations.

Averaging: Averaging the aligned subtomograms to generate a high-resolution 3D reconstruction of the structure. Averaging reduces noise and enhances the signal, resulting in a clearer depiction of the biological complex.

Applications of Subtomogram Averaging

Subtomogram averaging has had a profound impact on our understanding of cellular structures and macromolecular complexes. Some notable applications include:

Flagellar Motors in Bacteria: Subtomogram averaging has revealed the detailed architecture of bacterial flagellar motors, providing insights into their function and assembly.

Ribosomes: Investigating ribosome structure and organization within cells has been greatly enhanced by subtomogram averaging. It has allowed researchers to study ribosome heterogeneity and localization.

Mitochondrial Cristae: Understanding the inner mitochondrial membrane and its cristae morphology has been advanced through subtomogram averaging. It has implications for energy production and cellular health.

Viral Capsids: Subtomogram averaging has been used to examine the structure of viral capsids, shedding light on the assembly and maturation of viruses.

Challenges and Future Directions

While subtomogram averaging and tomography have revolutionized structural biology, they are not without challenges. The need for sophisticated instrumentation, extensive computational resources, and expertise in image processing can be barriers to entry. Moreover, radiation damage from electron beams remains a concern, limiting the resolution attainable for some samples.

Future developments in the field may focus on addressing these challenges. This could involve improving sample preparation techniques, enhancing data acquisition speed, and developing new algorithms for image reconstruction and subtomogram averaging. Additionally, innovations in cryo-tomography are likely to further expand our ability to study biological structures in their native, hydrated state.

Subtomogram averaging and tomography have ushered in a new era of structural biology, allowing researchers to explore cellular structures and macromolecular complexes with unprecedented detail and precision. As technology continues to advance and computational methods improve, these techniques will undoubtedly play an increasingly pivotal role in unravelling the mysteries of the cellular world, providing insights into health, disease, and the fundamental processes of life.

Chapter 12: Integrating Structural Biology Datasets

12.1 Approaches for combining X-ray, NMR, and cryo-EM data

The realm of structural biology has witnessed remarkable advances in recent decades, driven by innovations in experimental techniques and computational methodologies. While each of these techniques—X-ray crystallography, nuclear magnetic resonance (NMR) spectroscopy, and cryo-electron microscopy (cryo-EM)—offers valuable insights into biomolecular structures, they also possess their own limitations and constraints. As a result, researchers have increasingly turned to integrative approaches, combining data from multiple sources, to obtain a more comprehensive and accurate understanding of macromolecular structures. In this chapter, we explore the exciting world of integrative structural biology, focusing on the strategies and methods employed to harmoniously merge X-ray, NMR, and cryo-EM data.

The Motivation for Integration

Before delving into the intricacies of data integration, it is imperative to grasp why researchers are driven to combine these disparate structural biology techniques. Each of the three methods has its unique strengths and limitations, and these characteristics make them complementary to one another.

X-ray Crystallography offers high-resolution structural information but necessitates the formation of well-ordered crystals, which can be a major hurdle for some biomolecules. Moreover, it provides static snapshots of structures and may not capture dynamic aspects effectively.

NMR Spectroscopy excels in characterizing the dynamics of biomolecules in solution, offering insights into their conformational flexibility. However, it is limited in achieving high-resolution structures for larger complexes or proteins with high molecular weight.

Cryo-EM, while increasingly capable of achieving near-atomic resolution, can still face challenges in achieving atomic-level precision and defining side-chain conformations in complex structures.

By combining these techniques, researchers aim to harness the strengths of each while compensating for their limitations, thereby obtaining a more accurate and complete structural description of biological macromolecules. Here, we explore several approaches employed for this purpose.

Hybrid Methods

Hybrid methods represent one approach to integrating data from X-ray, NMR, and cryo-EM experiments. These methods involve the use of data from two or more of these techniques to improve structural accuracy and completeness. One common hybrid approach is the combination of X-ray crystallography and cryo-EM data, often referred to as X-ray-guided cryo-EM or cryo-EM-guided crystallography.

Example: Integrating X-ray and Cryo-EM Data for the Ribosome

A prominent example of hybrid methods involves the structural elucidation of the ribosome, a complex macromolecular assembly responsible for protein synthesis. The ribosome is composed of both protein and RNA components, making it challenging to crystallize and study using X-ray crystallography alone. Cryo-EM has been pivotal in obtaining lower-resolution structures of the ribosome. However, to achieve atomic-level detail, researchers combined high-resolution X-ray data from small ribosomal subunits with lower-resolution cryo-EM data from the larger subunits. This hybrid approach allowed for the creation of more complete and accurate ribosome structures, shedding light on its intricate mechanism of action.

Integrative Modelling

Integrative modelling is another powerful technique used to combine data from multiple sources. This method involves the simultaneous modelling of structures based on data from X-ray, NMR, and cryo-EM experiments, along with other biophysical and biochemical information. Integrative modelling approaches use computational algorithms to generate structural models consistent with all available data.

Example: Structural Determination of Membrane Proteins

Membrane proteins, which play critical roles in cellular function, often present significant challenges for structural characterization due to their hydrophobic nature. Integrative modelling has been instrumental in studying these proteins. Researchers collect NMR data on the protein in a membrane-mimicking environment and use this information, along with data from cryo-EM and X-ray experiments, to generate structural models. The integrative approach provides a more complete picture of the protein's structure, including its transmembrane regions and interactions with lipids.

Validation and Cross-Validation

Integrating data from multiple techniques requires rigorous validation to ensure the accuracy and reliability of the resulting structural models. Researchers employ various validation strategies, including cross-validation, to assess the consistency of data obtained from different methods.

Example: Cross-Validating NMR and Cryo-EM Data

In the structural characterization of protein complexes, researchers often use NMR to determine the structure of individual subunits and cryo-EM to study the overall complex. Cross-validation involves comparing the structural information obtained from NMR and cryo-EM to confirm their agreement. When the NMR-derived structures of individual subunits align with the cryo-EM-derived structures of the entire complex, it provides strong evidence for the accuracy of the integrated model.

Flexibility and Dynamics

One of the key advantages of integrating data from multiple techniques is the ability to capture the dynamic behaviour of biomolecules. Proteins and other macromolecules are rarely static; they undergo conformational changes that are crucial for their biological functions.

Example: Studying Protein Dynamics with NMR and Cryo-EM

NMR spectroscopy excels in providing insights into protein dynamics in solution. By collecting NMR data on a protein in various states and combining it with high-resolution cryo-EM data, researchers can create models that depict how the protein's structure changes as it performs its biological functions. This integrative approach has been invaluable in understanding processes such as enzyme catalysis and molecular recognition.

Software and Databases

To facilitate the integration of data from multiple sources, researchers have developed specialized software tools and databases. These resources streamline the process of data analysis, modelling, and validation.

Example: Integrative Structural Biology Software

Software packages like Integrative Modelling Platform (IMP) and HADDOCK provide researchers with the tools needed to perform integrative modelling and validation. IMP, for instance, allows for the incorporation of data from X-ray, NMR, and cryo-EM experiments, as well as other biophysical measurements, into a unified structural model. These platforms have become essential for researchers working in integrative structural biology.

The integration of data from X-ray crystallography, NMR spectroscopy, and cryo-EM represents a powerful approach in modern structural biology. By combining the strengths of these techniques, researchers can overcome individual limitations and gain a more comprehensive understanding of complex biomolecular structures. From hybrid methods to integrative modelling and validation strategies, the toolbox of integrative structural biology continues to expand, opening new avenues for unravelling the mysteries of the molecular world. This multidisciplinary approach holds great promise for advancing our knowledge of biology and accelerating drug discovery and development in the years to come.

12.2 Validating and cross-validating structural models

Validating and cross-validating structural models is a pivotal step in high-resolution protein structure determination. This process ensures the reliability and accuracy of the proposed atomic coordinates, which are the foundation of structural biology. In this section, we will explore the methods and strategies employed to rigorously validate protein structures, thereby enhancing our confidence in the derived models.

The Need for Validation

Protein structures generated through techniques such as X-ray crystallography, NMR spectroscopy, or cryo-electron microscopy provide a wealth of atomic information. However, these experimental methods are not immune to errors and limitations. Therefore, structural validation is indispensable to assess the quality and correctness of the derived models.

Common Validation Metrics

Several validation metrics are commonly employed to evaluate the quality of protein structures. Here, we will delve into some of the most widely used indicators:

Resolution: In X-ray crystallography and cryo-EM, resolution is a key metric. It quantifies the finest details that can be discerned in the electron density map. Higher resolution indicates a better-defined structure.

Example: A protein structure determined at a resolution of 1.5 Ångstroms provides more precise atomic positions than one at 3.0 Ångstroms.

R-Factor and R-Free: These metrics quantify the agreement between the observed and calculated data. R-factor assesses the fit between the experimental and calculated electron density, while R-free evaluates model overfitting by using a portion of the data not included during refinement.

Example: An R-factor of 0.18 indicates that 82% of the electron density is accounted for by the model.

Ramachandran Plot: This plot assesses the stereochemical quality of the protein backbone by analysing the dihedral angles of its constituent amino acids. It helps identify improbable or strained conformations.

Example: A Ramachandran plot should show most data points in the favoured and allowed regions, with only a small percentage in disallowed regions.

MolProbity Score: MolProbity is a comprehensive metric that evaluates several aspects of protein structure, including clashes between atoms, bond lengths, and angles. Lower MolProbity scores indicate better quality models.

Example: A MolProbity score of 1.0 is considered excellent, while higher scores indicate structural issues.

Cross-Validation: The Gold Standard

To enhance the robustness of structural models, cross-validation is employed. Cross-validation involves splitting the data into two subsets: one used for model building and the other for validation. The goal is to assess whether the model can predict the validation data accurately, effectively simulating an independent test of the model's quality.

Methods of Cross-Validation

Leave-One-Out Cross-Validation (LOOCV): In LOOCV, one data point (e.g., reflection in X-ray crystallography) is left out as the validation set, and the model is built using the remaining data points. This process is repeated iteratively, leaving out each point once.

Example: In X-ray crystallography, LOOCV involves omitting one reflection at a time to validate the model.

K-Fold Cross-Validation: K-fold cross-validation partitions the data into K subsets. The model is built using K-1 subsets and validated on the remaining subset. This process is repeated K times, with each subset serving as the validation set once.

Example: In cryo-EM, K-fold cross-validation might involve splitting particle images into K subsets for validation.

Bootstrapping: Bootstrapping is a resampling technique that involves randomly selecting data points with replacement to create multiple training and validation sets. It assesses the robustness of the model by generating a distribution of validation scores.

Example: In NMR spectroscopy, bootstrapping can be used to assess the precision of the derived structural parameters.

Cross-Validation Scores

Cross-validation provides scores or metrics that reflect the model's predictive accuracy on the validation data. Commonly used scores include:

R2 Score: The coefficient of determination measures the proportion of the variance in the validation data that is explained by the model. An $R2$ score close to 1 indicates a highly predictive model.

Example: An $R2$ score of 0.9 suggests that the model accounts for 90% of the variance in the validation data.

Cross-Validation Error: This metric quantifies the average error between the predicted values and the actual validation data.

Example: A low cross-validation error indicates that the model accurately predicts the validation data.

Integration of Validation Metrics

To ensure the highest confidence in a protein structure, multiple validation metrics should be considered collectively. A single validation metric, while informative, may not provide a comprehensive assessment of the model's quality. Therefore, an integrated approach is recommended, wherein multiple metrics are examined in conjunction.

Case Study: HIV-1 Protease Structure Validation

Let's examine a real-world example of how validation metrics were integrated in the determination of the HIV-1 protease structure. This enzyme plays a crucial role in the life cycle of the HIV virus and is a target for antiretroviral drugs.

In the case of HIV-1 protease, validation metrics such as resolution, R-factor, Ramachandran plot, and MolProbity score were meticulously assessed. The integration of these metrics provided a holistic view of the structural quality, ensuring that the derived model was accurate and reliable. This validation process was crucial in the development of effective HIV drugs.

Considering high-resolution protein structure determination, the validation and cross-validation of structural models are paramount. These processes serve as essential safeguards against errors and inaccuracies, providing researchers with confidence in the atomic coordinates of biological macromolecules. By diligently applying a suite of validation metrics and adopting cross-validation strategies, structural biologists ensure that their models stand up to rigorous scrutiny, advancing our understanding of protein function and enabling breakthroughs in drug design and therapeutic development. As structural biology continues to evolve, validation will remain a cornerstone of the field, reinforcing the integrity of the structural data that underpins our understanding of life at the molecular level.

Chapter 13: Membrane Protein Structure Determination

13.1 Challenges and techniques for membrane protein structures

Membrane proteins play a pivotal role in various cellular processes, serving as gatekeepers for molecular transport and as signal transduction hubs. Understanding their structures is of paramount importance, as it can shed light on their functions and enable drug development targeting these proteins. However, the structural determination of membrane proteins poses unique challenges compared to their soluble counterparts. In this subsection, we will delve into the intricacies and difficulties associated with studying membrane protein structures, along with the innovative techniques that have been developed to overcome these hurdles.

The Hydrophobic Dilemma

Membrane proteins are predominantly hydrophobic in nature due to their integral association with the lipid bilayer of cell membranes. This inherent hydrophobicity presents a significant challenge during the purification and crystallization stages. Unlike soluble proteins, which can be easily solubilized in aqueous solutions, membrane proteins require detergents or lipid environments to maintain their structural integrity.

One of the key techniques employed to address this challenge is the use of detergents. Detergents solubilize membrane proteins by encapsulating their hydrophobic regions, mimicking the lipid environment. However, selecting the appropriate detergent is critical, as the wrong choice can disrupt the native conformation of the protein or lead to aggregation. Moreover, the detergent micelles themselves can interfere with crystal packing in X-ray crystallography, making crystallization more challenging.

Example 1: Detergent Selection for Membrane Protein Crystallization

In a study published in the journal Nature, researchers tackled the detergent selection problem when crystallizing the membrane protein bacteriorhodopsin. They systematically tested a panel of detergents and found that the choice of detergent dramatically influenced the quality of crystals obtained. This research highlighted the importance of detergent screening to optimize crystallization conditions for membrane proteins.

Example 2: Lipid Nanodiscs

To address the issues associated with detergents, researchers have developed lipid nanodiscs. These nanodiscs consist of a lipid bilayer stabilized by membrane scaffold proteins. Membrane proteins can be reconstituted into these nanodiscs, providing a more native-like lipid environment for structural studies. This innovative approach has led to successful structural determinations of numerous membrane proteins.

Overcoming Conformational Heterogeneity

Another significant challenge in membrane protein structure determination is the inherent flexibility and conformational heterogeneity of these proteins. Membrane proteins often exist in multiple conformational states, which can complicate crystallization and structure determination.

Example 3: The G Protein-Coupled Receptor (GPCR) Conundrum

GPCRs are a class of membrane proteins known for their involvement in signal transduction. These proteins are notorious for their conformational flexibility, switching between active and inactive states. Structural studies of GPCRs were historically challenging due to this flexibility. However, recent advances in cryo-electron microscopy (cryo-EM) have enabled researchers to capture multiple conformational states of GPCRs, providing crucial insights into their mechanisms of action.

Techniques for Stabilization and Crystallization

To address conformational heterogeneity and flexibility, researchers have developed techniques for stabilizing membrane proteins in specific states. One such approach is the use of antibody fragments or nanobodies that bind to and stabilize a particular conformation of the protein. This technique, known as "conformational trapping," has been instrumental in capturing previously elusive membrane protein structures.

Example 4: Conformational Trapping of Aquaporin

In a groundbreaking study published in Science, scientists used conformation-specific nanobodies to trap and crystallize the aquaporin-2 water channel in an open conformation. This allowed for the determination of the first high-resolution structure of an open-state aquaporin, providing valuable insights into water transport mechanisms.

Innovations in Electron Microscopy: Cryo-EM

Cryo-electron microscopy (cryo-EM) has emerged as a powerful technique for studying membrane proteins, particularly those resistant to crystallization. Cryo-EM allows for the imaging of proteins in their near-native state, embedded in a thin layer of vitrified ice. This method has the advantage of not requiring crystallization and can capture protein conformations in a more physiologically relevant environment.

Example 5: Structure of the Mitochondrial Respiratory Supercomplex

A study published in Nature Communications employed cryo-EM to elucidate the structure of the mitochondrial respiratory supercomplex, a large and challenging membrane protein assembly. This research revealed the intricate arrangement of protein subunits within the supercomplex and provided insights into cellular energy production.

Hybrid Methods for Membrane Protein Structure Determination

Hybrid methods that combine data from different structural techniques have gained prominence in membrane protein research. Integrating data from X-ray crystallography, NMR spectroscopy, and cryo-EM can provide a more comprehensive view of a membrane protein's structure and dynamics.

Example 6: The Structural Characterization of the ABC Transporter

A study published in Science Advances integrated X-ray crystallography, cryo-EM, and NMR spectroscopy to determine the structure of an ABC transporter, a class of membrane proteins involved in the transport of various molecules across cell membranes. This hybrid approach allowed researchers to capture the transporter in different functional states and provided insights into its transport mechanism.

Studying membrane protein structures is a challenging but crucial endeavour in structural biology. The unique hydrophobic nature, conformational flexibility, and inherent heterogeneity of membrane proteins demand innovative techniques and approaches for successful structure determination. As technology continues to advance, researchers are poised to unlock the mysteries of these vital cellular components, paving the way for new drug targets and a deeper understanding of cellular function.

13.2 Detergents and lipid nanodiscs in membrane protein preparation

The structural elucidation of membrane proteins, which are integral components of cell membranes and play vital roles in various biological processes, has long posed a significant challenge in structural biology. These proteins are notorious for their hydrophobic nature, making their extraction, purification, and crystallization a complex endeavour. To overcome these challenges, researchers have developed innovative approaches, including the use of detergents and lipid nanodiscs. In this section, we will delve into the intricacies of these two methods, exploring their applications, advantages, and limitations in preparing membrane proteins for high-resolution structural studies.

The Challenge of Membrane Proteins

Membrane proteins are a diverse group that includes receptors, transporters, and enzymes, and they are crucial for mediating cellular interactions and signalling processes. However, their unique structural and biochemical properties have made them elusive targets for structural biologists. One of the key challenges in studying membrane proteins is their hydrophobic transmembrane regions, which are incompatible with the aqueous environments required for traditional biochemical techniques.

The Role of Detergents

Detergents: Amphiphilic Molecules

Detergents are amphiphilic molecules that contain both hydrophilic and hydrophobic domains. This dual nature allows them to solubilize hydrophobic membrane proteins by encapsulating their hydrophobic regions within micelles, where the hydrophobic tails of the detergent molecules shield the hydrophobic segments of the protein, while the hydrophilic heads interact with water molecules, thus rendering the protein soluble in an aqueous medium.

Detergent Selection

The choice of detergent is critical in membrane protein preparation. Researchers must consider factors such as the size of the hydrophobic domain of the protein, the protein's stability, and the specific requirements of downstream experiments. Commonly used detergents for membrane protein solubilization include dodecyl maltoside (DDM), n-dodecyl-β-D-maltopyranoside (DM), and lauryl dimethylamine oxide (LDAO). These detergents have varying properties, and their compatibility with a particular membrane protein depends on its unique characteristics.

Protein Extraction and Solubilization

To begin the membrane protein preparation process, cell membranes are typically isolated from the organism or cells of interest. The membranes are then solubilized with an appropriate detergent. During this step, it is essential to strike a balance between achieving complete solubilization of the target protein and preventing excessive denaturation or aggregation. Optimization of detergent concentration and buffer conditions is often necessary to achieve these goals.

Example 1: Bacteriorhodopsin

One of the earliest successes in membrane protein crystallization using detergents was the structure determination of bacteriorhodopsin, a light-driven proton pump found in halophilic archaea. In this case, the detergent used was octylglucoside, which facilitated the extraction and stabilization of bacteriorhodopsin, ultimately leading to the first crystal structure of a membrane protein in 1985.

Detergent Exchange

After solubilization, researchers often perform detergent exchange to replace the initial detergent with one more suitable for downstream applications, such as crystallization or NMR studies. This step is crucial because the original detergent may interfere with the stability of the protein or the quality of the crystals.

Example 2: G protein-coupled receptors (GPCRs)

The study of GPCRs, a prominent class of membrane proteins involved in cell signalling, exemplifies the significance of detergent exchange. Detergents such as octylthioglucoside (OTG) or n-decyl-β-D-maltopyranoside (DM) have been used to solubilize GPCRs, but for structural studies, it is often necessary to exchange these detergents with lipids or lipid-like molecules.

Detergent Removal

In some cases, detergents may be completely removed from the system, particularly when pursuing NMR studies, where detergents can interfere with spectral quality. Techniques like dialysis, gel filtration, or the use of specialized detergents with removable tags (e.g., maltose-binding protein fusions) are employed to eliminate detergents while preserving the native state of the protein.

Challenges with Detergents

While detergents have been instrumental in the solubilization and stabilization of membrane proteins, they also have limitations. One significant challenge is their potential to disrupt the native lipid environment of the protein, which can influence its structure and function. Additionally, some detergents may not work well with certain membrane proteins, leading to denaturation or aggregation.

The Advent of Lipid Nanodiscs

Lipid Nanodiscs: Maintaining Native Lipid Environment

Lipid nanodiscs represent a revolutionary approach to overcome the limitations associated with detergents. These nanoscale lipid bilayers, stabilized by membrane scaffold proteins (MSPs), provide a means to incorporate membrane proteins while maintaining their native lipid environment.

Formation of Lipid Nanodiscs

Lipid nanodiscs are formed by mixing lipids, a membrane scaffold protein (MSP), and the membrane protein of interest in an aqueous solution. The MSP self-assembles around the lipids, encapsulating them in a bilayer structure. The membrane protein can then be incorporated into the lipid bilayer of the nanodisc.

Example 3: Rhodopsin

The structure determination of rhodopsin, a prototypical GPCR, is a notable example of the successful application of lipid nanodiscs. By incorporating rhodopsin into nanodiscs composed of native retinal lipids, researchers were able to obtain a structure that closely resembled the protein's native state, highlighting the advantages of preserving the native lipid environment.

Advantages of Lipid Nanodiscs

Lipid nanodiscs offer several advantages over traditional detergent-based methods:

Native Lipid Environment: Lipid nanodiscs maintain the natural lipid composition surrounding the membrane protein, which can be critical for maintaining its stability and function.

Stoichiometry Control: Nanodiscs provide precise control over the ratio of lipid to protein, ensuring the desired stoichiometry.

Compatibility with Multiple Techniques: Nanodiscs are compatible with various structural and biophysical techniques, including NMR, cryo-EM, and X-ray crystallography.

Improved Stability: The lipid bilayer in nanodiscs can enhance the stability of membrane proteins, reducing denaturation and aggregation.

Challenges with Lipid Nanodiscs

Despite their advantages, lipid nanodiscs are not without challenges. Their assembly and optimization can be labour-intensive, and the choice of MSP and lipid composition must be carefully considered. Additionally, the size of nanodiscs can affect their behaviour in downstream experiments, necessitating selection based on the specific requirements of the study.

In the quest to decipher the structures and functions of membrane proteins, the use of detergents and lipid nanodiscs represents two pivotal approaches. Detergents have been instrumental in solubilizing and stabilizing membrane proteins, allowing for their characterization through various structural techniques. On the other hand, lipid nanodiscs have emerged as a game-changing technology, preserving the native lipid environment and providing a versatile platform for structural studies.

The choice between detergents and lipid nanodiscs depends on the specific goals of the study, the nature of the membrane protein, and the intended downstream applications. Researchers continue to refine these methods, pushing the boundaries of our understanding of membrane protein structure and function, ultimately contributing to advancements in drug discovery and biotechnology.

13.3 *Recent advances in membrane protein structure determination*

Membrane proteins represent a vital class of biomolecules that play pivotal roles in numerous physiological processes, making them significant targets for drug discovery and therapeutic interventions. However, their hydrophobic nature and inherent instability have long posed formidable challenges to structural biologists. In this subsection, we explore the recent and groundbreaking advances in membrane protein structure determination, highlighting innovative techniques, novel approaches, and notable achievements that have revolutionized our understanding of these essential cellular components.

Cryo-Electron Microscopy (Cryo-EM): A Game-Changer for Membrane Proteins

One of the most remarkable breakthroughs in structural biology, especially in the context of membrane proteins, has been the widespread adoption of cryo-electron microscopy (Cryo-EM). Historically, membrane protein crystallization was fraught with difficulties due to their preference for the lipid bilayer environment. Cryo-EM, however, bypasses the need for crystallization, allowing the study of membrane proteins in their native lipid membranes.

In recent years, Cryo-EM has been instrumental in unveiling the structures of challenging membrane proteins that had long resisted crystallization. The 2017 Nobel Prize in Chemistry recognized this transformative technology, with Dr. Jacques Dubochet, Dr. Joachim Frank, and Dr. Richard Henderson being awarded for their pioneering contributions to Cryo-EM.

A striking example of Cryo-EM's success lies in the elucidation of the structure of the TRPV1 ion channel, which is involved in pain sensation and temperature regulation. Traditional methods struggled to capture the dynamic behaviour of this protein. However, in 2016, researchers used Cryo-EM to reveal the TRPV1 structure in multiple conformations, shedding light on its gating mechanism.

Advancements in Sample Preparation for Cryo-EM

Sample preparation has been a bottleneck in Cryo-EM studies, particularly for membrane proteins. Recent developments in this area have significantly improved the quality and throughput of data acquisition.

One notable advancement is the use of lipid nanodiscs, which provide a stable, native-like environment for membrane proteins. By incorporating the target protein into these nanodiscs, researchers can maintain the protein's structural integrity, allowing for high-resolution imaging. For instance, the structure of the β2-adrenergic receptor, a G-protein-coupled receptor involved in cell signalling, was elucidated using Cryo-EM with the assistance of lipid nanodiscs.

Moreover, innovations in detergent screening and the development of amphipols, which are amphiphilic polymers that can encapsulate membrane proteins, have expanded the toolkit for membrane protein sample preparation. These advances have proven invaluable in obtaining well-diffracting images.

MicroED: A Potential Complement to Cryo-EM

While Cryo-EM has made remarkable strides in membrane protein structure determination, it is essential to explore complementary techniques to tackle challenging cases. Microcrystal electron diffraction (MicroED) has emerged as a promising approach for membrane protein crystallography, especially for small and radiation-sensitive crystals.

MicroED involves collecting electron diffraction data from nanometre-sized crystals, which can be advantageous for membrane proteins, as they often form small and fragile crystals. Notably, MicroED has been used to determine the structure of bacteriorhodopsin, a prototypical membrane protein, at high resolution.

This technique's power lies in its ability to exploit minute crystals that may not yield usable data with traditional X-ray crystallography methods. Thus, MicroED offers a complementary avenue for studying membrane proteins when Cryo-EM or NMR spectroscopy may not be applicable.

Combining Cryo-EM with NMR Spectroscopy: Hybrid Approaches

Hybrid methods that combine data from multiple structural techniques have gained traction in recent years, offering an integrated approach to tackle membrane protein structural challenges. In particular, the combination of Cryo-EM and nuclear magnetic resonance (NMR) spectroscopy has shown promise in providing both high-resolution structures and dynamic insights.

For example, the structure of the membrane protein EmrE, a multidrug transporter, was determined by combining Cryo-EM and solid-state NMR. This approach allowed researchers to gain a comprehensive understanding of the protein's structure and dynamics in a lipid bilayer environment.

Additionally, advancements in isotope labelling and magic-angle spinning NMR have facilitated the study of membrane protein dynamics and interactions, complementing the static structural information obtained from Cryo-EM. Integrative modelling techniques have emerged to merge data from different sources, enabling researchers to generate hybrid models that capture both static and dynamic aspects of membrane proteins.

Emerging Technologies: Serial Crystallography and Free-Electron Lasers

Looking to the future, emerging technologies hold great promise for membrane protein structure determination. Serial crystallography, which involves collecting diffraction data from a series of microcrystals, has gained attention for its potential to overcome challenges associated with membrane protein crystallization. Free-electron lasers (FELs) have further enhanced serial crystallography by providing ultrabright X-ray sources, enabling data collection from smaller crystals with minimal radiation damage.

Serial femtosecond crystallography (SFX), a form of serial crystallography using FELs, has been used to determine the structure of membrane proteins like the photosynthetic reaction centre and the photosystem II complex. These advances promise to expand our ability to study membrane proteins under more physiologically relevant conditions.

In Silico Approaches and AI-Assisted Modelling

In silico approaches, aided by artificial intelligence (AI) and machine learning, are increasingly playing a pivotal role in membrane protein structure determination. AI algorithms are being employed to predict protein structures and model their interactions with ligands and lipid bilayers.

For instance, AlphaFold, developed by DeepMind, demonstrated remarkable accuracy in predicting protein structures, including membrane proteins, from their amino acid sequences. Such AI-driven advances have the potential to accelerate membrane protein structure determination by providing initial structural models for further refinement and validation.

Concluding Remarks

The recent advances in membrane protein structure determination have opened new frontiers in our understanding of these critical biomolecules. Cryo-EM has emerged as a transformative technique, while innovations in sample preparation, hybrid methods, and emerging technologies continue to push the boundaries of what is achievable in this field. In silico approaches, powered by AI, are poised to further streamline the process of membrane protein structure determination.

As the membrane protein structural landscape continues to evolve, these advances not only shed light on the molecular mechanisms underlying essential biological processes but also hold immense potential for drug discovery and the development of targeted therapies. With ongoing research and technological innovations, we can anticipate even more exciting breakthroughs in the near future, expanding our knowledge of membrane proteins and their roles in health and disease.

Chapter 14: Protein-Ligand Interactions

14.1 Studying protein-ligand complexes using high-resolution techniques

Protein-ligand complexes lie at the heart of molecular interactions that govern various biological processes, making them a focal point in drug discovery, structural biology, and biochemistry. High-resolution techniques have revolutionized the study of these complexes, shedding light on the intricate binding mechanisms, energetics, and functional consequences. In this section, we will delve into the captivating world of protein-ligand interactions, exploring how high-resolution methods such as X-ray crystallography, NMR spectroscopy, and cryo-electron microscopy (cryo-EM) enable us to unravel the mysteries of molecular recognition.

X-ray Crystallography: Peering into the Atomic Landscape

X-ray crystallography has been a cornerstone in deciphering the atomic details of protein-ligand complexes. By forming well-ordered crystals of the protein-ligand complex, researchers can direct a beam of X-rays onto the crystal, producing a diffraction pattern that is then transformed into a three-dimensional electron density map of the complex. This map reveals the precise arrangement of atoms, including those of the ligand and the protein, with resolutions typically reaching the sub-angstrom level (0.1 nm).

One remarkable example of X-ray crystallography's power is the study of the complex between the enzyme acetylcholinesterase (AChE) and the nerve agent sarin. AChE is essential for the regulation of neurotransmission by catalyzing the hydrolysis of the neurotransmitter acetylcholine. Sarin, a potent organophosphorus compound, irreversibly inhibits AChE, leading to toxic effects on the nervous system. Through X-ray crystallography, researchers elucidated the atomic-level interactions between sarin and AChE. They discovered that sarin covalently binds to a serine residue in the enzyme's active site, effectively disabling its catalytic function. This structural insight has guided the development of antidotes against nerve agents and enhanced our understanding of enzymatic inhibition.

NMR Spectroscopy: Capturing Dynamics in Binding

Nuclear magnetic resonance (NMR) spectroscopy offers a unique perspective on protein-ligand complexes by unveiling not only their static structures but also their dynamic behavior in solution. NMR detects the nuclear spin properties of atoms, providing information about distances, angles, and motions of individual atoms within the complex. This versatility allows researchers to study a wide range of interactions, including weak and transient binding events.

Consider the case of the interaction between the HIV-1 protease enzyme and its inhibitor, darunavir. High-resolution NMR spectroscopy revealed the binding mechanism of darunavir to the protease, showcasing the role of conformational dynamics in the binding process. The study demonstrated that the binding pocket of the protease is highly flexible, accommodating the inhibitor through a series of dynamic structural changes. This flexibility is a key feature in understanding the resistance of HIV to various drugs and has led to the design of more effective protease inhibitors.

Cryo-EM: Visualizing Large and Complex Complexes

Cryo-electron microscopy (cryo-EM) has emerged as a powerful technique for visualizing protein-ligand complexes, particularly those that are large, heterogeneous, or challenging to crystallize. Cryo-EM involves freezing the sample in vitreous ice and then imaging it with an electron microscope. Recent advances in detector technology and image processing have pushed the resolution limits of cryo-EM, allowing researchers to visualize protein-ligand complexes at near-atomic resolutions.

One of the most remarkable cryo-EM studies elucidated the structure of the ribosome in complex with antibiotics. The ribosome is a massive macromolecular complex responsible for protein synthesis, making it a vital target for antibiotics. By visualizing the ribosome in complex with antibiotics like erythromycin and tetracycline, cryo-EM provided insights into how these drugs disrupt bacterial protein synthesis. This knowledge has informed the development of new antibiotics and strategies to combat antibiotic resistance.

Binding Affinities and Thermodynamics

High-resolution techniques not only reveal the geometry of protein-ligand complexes but also provide essential information about the thermodynamics of binding. One can determine the binding affinity (Kd) of a ligand for its target protein, which quantifies the strength of the interaction. This parameter is crucial for understanding the effectiveness of a drug or the specificity of a ligand.

For instance, X-ray crystallography can reveal the structural basis of ligand binding, allowing researchers to identify key interactions that contribute to binding affinity. By analysing the electron density maps and intermolecular contacts, it becomes possible to design ligands with improved binding affinities or selectivity.

NMR spectroscopy can complement this information by measuring the changes in chemical shifts or relaxation rates of atoms in the protein upon ligand binding. These changes provide insights into the thermodynamic parameters such as enthalpy (ΔH) and entropy (ΔS) changes associated with binding. Understanding these thermodynamic factors aids in the rational design of ligands with optimal binding properties.

Allosteric Modulation: Beyond the Active Site

High-resolution techniques have also unveiled the fascinating world of allosteric modulation, where ligands bind to sites on a protein distant from the active site, influencing its activity. Allosteric regulation plays a crucial role in many biological processes and is a target for drug discovery.

An illustrative example is the binding of allosteric modulators to G-protein-coupled receptors (GPCRs). GPCRs are membrane proteins that play key roles in signal transduction. Allosteric modulators can bind to distinct sites on GPCRs, altering their conformation and affecting downstream signalling. The crystal structures of GPCRs bound to allosteric modulators have provided crucial insights into the allosteric mechanisms, paving the way for the development of drugs with improved specificity and reduced side effects.

Emerging Techniques and Future Prospects

High-resolution techniques continue to evolve, offering exciting prospects for studying protein-ligand complexes. For instance, advancements in serial femtosecond crystallography (SFX) have enabled the study of extremely radiation-sensitive samples, including membrane proteins and transient intermediates. In combination with X-ray free-electron lasers (XFELs), SFX has the potential to capture dynamic processes such as enzymatic reactions in real-time.

Moreover, the integration of computational approaches and artificial intelligence (AI) is accelerating the analysis and interpretation of high-resolution data. Machine learning algorithms can predict binding affinities, identify potential ligand-binding sites, and facilitate the de novo design of ligands, opening new avenues for drug discovery.

High-resolution techniques have transformed our understanding of protein-ligand interactions. From revealing atomic structures to elucidating binding thermodynamics and uncovering allosteric regulation, these techniques have broadened our knowledge of molecular recognition. As technology continues to advance, the future promises even deeper insights into the complex dance of proteins and ligands, with profound implications for drug discovery and our understanding of biology.

14.2 Structure-based drug design

Structure-based drug design (SBDD) is a powerful approach in the field of drug discovery that utilizes high-resolution protein structures to design novel therapeutics with enhanced specificity and efficacy. This subsection explores the principles of SBDD, its applications, success stories, and the impact of structural biology on the pharmaceutical industry.

Principles of Structure-based Drug Design

Structure-based drug design begins with the determination of the three-dimensional structure of the target protein, typically using techniques such as X-ray crystallography, NMR spectroscopy, or cryo-electron microscopy (cryo-EM). Once the protein structure is known, computational tools and molecular modelling techniques are employed to identify potential drug-binding sites and design small molecules or ligands that can interact with the target protein.

Example: In the case of the HIV protease inhibitor, Darunavir, the crystal structure of the HIV protease enzyme was determined, revealing its active site. Researchers used this structural information to design a potent inhibitor, Darunavir, which specifically binds to the active site, preventing viral replication.

Applications of Structure-based Drug Design

SBDD has found applications in various disease areas, including cancer, infectious diseases, and neurodegenerative disorders. The ability to rationally design molecules that interact with specific proteins has accelerated drug discovery and development processes.

Example: Imatinib, a breakthrough drug for chronic myeloid leukaemia (CML), was developed using SBDD. The drug precisely targets the BCR-ABL fusion protein, which drives CML. By designing a molecule that fits into the active site of this aberrant protein, Imatinib effectively inhibits its activity, leading to disease remission in many patients.

Success Stories in Structure-based Drug Design

Several drugs developed through SBDD have made significant contributions to medicine and patient care.

Tamiflu (Oseltamivir): In the case of influenza, the viral neuraminidase enzyme was the target of SBDD. The structure of neuraminidase provided insights into its function, enabling the design of Tamiflu, which blocks the enzyme's activity and is used to treat and prevent flu infections.

Relenza (Zanamivir): Similar to Tamiflu, Relenza was designed using SBDD to target neuraminidase. This inhaled drug is another effective treatment for influenza.

Protease Inhibitors for HIV: As mentioned earlier, HIV protease inhibitors like Darunavir and Saquinavir were developed using SBDD. They are vital components of highly active antiretroviral therapy (HAART) for HIV/AIDS patients.

Alectinib (Alecensa): Alectinib, used to treat non-small cell lung cancer (NSCLC), targets the anaplastic lymphoma kinase (ALK) protein. SBDD helped in the design of Alectinib to fit precisely into the ALK kinase domain, inhibiting its activity and slowing cancer progression.

These examples illustrate the transformative impact of SBDD in the pharmaceutical industry, where the precise targeting of disease-associated proteins has led to the development of safer and more effective drugs.

Challenges in Structure-based Drug Design

While SBDD has delivered remarkable successes, it also faces several challenges:

Structural Diversity: Not all drug targets are amenable to structural determination, and obtaining high-quality structures can be challenging for certain proteins.

Flexibility: Proteins are dynamic entities, and their structures can change upon ligand binding. Accounting for this flexibility is a complex task in SBDD.

Resistance: As with any therapeutic approach, drug resistance can develop, requiring constant innovation in the design of new drugs.

Off-Target Effects: Overly specific drug designs may lead to unexpected off-target effects, emphasizing the need for a balance between specificity and broader applicability.

Future Directions in Structure-based Drug Design

Advancements in structural biology and computational modelling are driving the future of SBDD:

Integration of Multiple Techniques: Combining data from X-ray crystallography, NMR, and cryo-EM can provide a more comprehensive understanding of protein-ligand interactions.

Machine Learning and AI: AI algorithms are being used to analyse large datasets and predict potential drug-binding sites, accelerating the early stages of drug discovery.

Fragment-based Drug Design: This approach involves designing smaller, fragment-sized molecules that bind to a protein. These fragments can then be chemically optimized to create more potent drugs.

Personalized Medicine: SBDD allows for the design of drugs tailored to individual genetic variations, leading to more effective and personalized treatments.

Impact on the Pharmaceutical Industry

The integration of SBDD into drug discovery pipelines has had a profound impact on the pharmaceutical industry. It has reduced the attrition rate of drug candidates, minimized side effects, and accelerated the development of novel therapies. The ability to design drugs with a high degree of specificity has revolutionized the treatment of many diseases, from cancer to infectious diseases.

Structure-based drug design represents a remarkable marriage of biology, chemistry, and computational science. It has transformed drug discovery by allowing researchers to design molecules with precision, targeting specific disease-related proteins. The success stories of drugs like Tamiflu, Imatinib, and HIV protease inhibitors highlight the effectiveness of SBDD in addressing some of the world's most challenging medical conditions. As technology continues to advance, the future of

SBDD holds promise for even more innovative and personalized therapies, ultimately improving patient care and the pharmaceutical industry as a whole.

14.3 Binding kinetics and thermodynamics

In a complex world of molecular interactions, understanding the dynamics of binding events is paramount. In this chapter, we investigate into the realm of binding kinetics and thermodynamics, revealing the intricate thermodynamic forces at play and the intricacies of binding kinetics that dictate the rate of association and dissociation between proteins and their ligands.

Binding Kinetics: The Molecular Choreography

Protein-ligand interactions are like a finely choreographed dance, where the partners—proteins and ligands—come together and separate with precise timing. This choreography is governed by binding kinetics, which comprises two crucial steps: association and dissociation.

Association Rate (k_on): The association rate, often denoted as k_on, represents the speed at which a ligand binds to a protein. It quantifies the likelihood of a collision between the two molecules leading to a productive interaction. Mathematically, k_on is expressed as:

kon = (Rate of association) / [Protein][Ligand]

Here, the rate of association is the number of successful binding events per unit time, and [Protein] and [Ligand] are the concentrations of the protein and ligand, respectively.

Dissociation Rate (k_off): The dissociation rate, k_off, measures how quickly the bound complex falls apart, reverting the protein and ligand to their unbound states. It is expressed as:

koff = Rate of dissociation / [Complex]

Where the rate of dissociation is the number of dissociation events per unit time, and [Complex] is the concentration of the bound complex.

Equilibrium Dissociation Constant (K_d): The equilibrium dissociation constant, denoted as K_d, is a fundamental parameter in binding kinetics. It quantifies the balance between association and dissociation and is calculated as the ratio of k_off to k_on:

Kd=kon/koff

A lower K_d indicates a tighter binding affinity, as it signifies a slower dissociation rate compared to the association rate.

Example: Haemoglobin and Oxygen

Consider the classic example of haemoglobin and oxygen. Haemoglobin (Hb) binds oxygen (O_2) in the lungs and releases it in tissues, facilitating oxygen transport in the bloodstream. The binding kinetics of this interaction are crucial for its physiological function.

Hemoglobin's k_on for oxygen binding is high, ensuring rapid association when oxygen is available. However, its k_off is also relatively high, enabling efficient oxygen release in tissues. The equilibrium dissociation constant (K_d) for the Hb-O_2 interaction is low, around 0.1-0.3 µM, indicating strong binding and efficient oxygen loading and unloading.

Thermodynamics of Binding: The Energetic Landscape

Understanding binding kinetics is only part of the story. The thermodynamics of binding reveal the energy changes associated with protein-ligand interactions. Key thermodynamic parameters include enthalpy (ΔH), entropy (ΔS), and Gibbs free energy (ΔG).

Enthalpy (ΔH): Enthalpy represents the heat exchange during binding. A negative ΔH indicates that binding is exothermic, releasing heat, while a positive ΔH suggests an endothermic process, absorbing heat. Enthalpy can be determined using techniques like isothermal titration calorimetry (ITC).

Entropy (ΔS): Entropy quantifies the disorder or randomness in a system. A positive ΔS indicates an increase in randomness upon binding, while a negative ΔS suggests a decrease. Changes in entropy are often associated with changes in conformation or solvation.

Gibbs Free Energy (ΔG): The Gibbs free energy change, ΔG, combines enthalpy and entropy to determine whether a binding event is thermodynamically favourable. It is calculated using the equation:

$$\Delta G = \Delta H - T\Delta S$$

Where T is the absolute temperature in Kelvin. If ΔG is negative, the binding is spontaneous and favourable. Conversely, a positive ΔG indicates an unfavourable binding event.

Example: Enzyme-Substrate Interactions

Enzyme-substrate interactions exemplify the thermodynamics of binding. Enzymes catalyse biochemical reactions by binding to substrates. The binding of a substrate to an enzyme often involves enthalpic and entropic changes.

The binding of a substrate to an enzyme is typically exothermic ($\Delta H < 0$) due to the formation of favourable non-covalent interactions, such as hydrogen bonds and hydrophobic interactions. However, there is often an entropic cost ($\Delta S < 0$) associated with substrate binding, as the substrates become more ordered upon binding to the enzyme's active site.

The net Gibbs free energy change (ΔG) determines whether the enzyme-substrate interaction is thermodynamically favourable. If ΔG is negative, the binding is spontaneous, and the enzyme catalyses the reaction efficiently.

Experimental Techniques for Binding Kinetics and Thermodynamics

Several experimental techniques are employed to measure binding kinetics and thermodynamics, providing valuable insights into protein-ligand interactions:

Isothermal Titration Calorimetry (ITC): ITC directly measures enthalpy changes during binding, allowing the determination of ΔH and K_d.

Surface Plasmon Resonance (SPR): SPR measures changes in refractive index upon binding, enabling the determination of k_{on}, k_{off}, and K_d.

Fluorescence Spectroscopy: Fluorescently labelled proteins or ligands are used to monitor binding-induced changes in fluorescence, yielding kinetic and thermodynamic information.

NMR Relaxation Methods: NMR spectroscopy can provide insights into binding kinetics by monitoring changes in relaxation rates of nuclei in the binding interface.

Steady-State and Transient Kinetics: Enzyme-substrate interactions are often studied using techniques like stopped-flow and rapid-quench kinetics to measure rates of association and dissociation.

Understanding binding kinetics and thermodynamics is pivotal in unravelling the intricate world of protein-ligand interactions. These parameters not only shed light on the molecular choreography of binding but also provide insights into the thermodynamic forces that drive these interactions. From the dance of haemoglobin with oxygen to the catalytic prowess of enzymes, the study of binding kinetics and thermodynamics continues to unlock the secrets of the molecular world.

In the following chapters, we will explore practical applications of these principles in drug discovery, structural biology, and beyond, showcasing how a deep understanding of binding kinetics and thermodynamics can impact various fields of science and medicine.

Chapter 15: Structural Studies of Protein Dynamics

15.1 Dynamics in protein structures

Proteins, the workhorses of biological systems, owe their functionality not just to their static three-dimensional conformations but also to their dynamic behaviour. While high-resolution techniques like X-ray crystallography, NMR spectroscopy, and cryo-electron microscopy (cryo-EM) provide invaluable snapshots of protein structures, they often fall short in capturing the full complexity of protein dynamics. In this section, we delve into the fascinating world of protein dynamics, exploring why they matter, the techniques used to study them, and their implications in understanding protein function.

Why Protein Dynamics Matter?

Proteins are not static entities; they constantly undergo dynamic motions on various timescales, from femtoseconds to seconds and beyond. These motions are not mere fluctuations but serve essential biological functions. Understanding protein dynamics is crucial for several reasons:

Enzyme Catalysis: Many enzymes rely on dynamic conformational changes to catalyse chemical reactions. A classic example is the enzyme adenylate kinase, where the closing and opening of specific domains are integral to its catalytic cycle. In such cases, a static structure alone cannot explain the enzyme's catalytic mechanism.

Ligand Binding: Proteins often undergo structural changes upon ligand binding. This ligand-induced fit is a dynamic process crucial for molecular recognition. The classic example is haemoglobin, which undergoes a conformational change upon oxygen binding, increasing its affinity for subsequent oxygen molecules.

Allosteric Regulation: Allosteric proteins are masters of dynamics. Conformational changes in one part of the protein can influence the function of distant sites. Understanding these dynamic communication pathways is vital for drug design and therapeutic interventions.

Protein Folding: Protein folding is an intricate dance of dynamic events. The energy landscape of a protein guides it through numerous conformations before reaching its native state. Misfolding and aggregation are associated with several diseases, making the study of dynamics critical.

Signalling Pathways: Signal transduction in cells relies heavily on protein dynamics. Receptor proteins change conformation upon ligand binding, setting off a cascade of intracellular events. Dysregulation of these dynamics can lead to diseases like cancer.

Techniques for Studying Protein Dynamics

Studying protein dynamics poses unique challenges. Unlike static structures, dynamics involve a range of motions, from side-chain fluctuations to backbone motions, domain movements, and global conformational changes. Researchers have developed several techniques to probe these dynamics:

Nuclear Magnetic Resonance (NMR) Spectroscopy: NMR is a powerful tool for studying protein dynamics in solution. It can provide insights into timescales ranging from picoseconds to milliseconds. Researchers use various NMR parameters, such as relaxation rates and chemical shift changes, to infer dynamic behaviour.

For example, NMR studies on protein G B1 domain revealed picosecond-to-nanosecond timescale motions of side chains. These motions were essential for ligand binding and molecular recognition.

Molecular Dynamics (MD) Simulations: MD simulations leverage the principles of classical mechanics to simulate protein dynamics at atomic resolution over extended timescales, often up to microseconds or longer. Researchers use these simulations to study a wide range of dynamic processes.

In a notable study, MD simulations of the enzyme dihydrofolate reductase (DHFR) elucidated the dynamic conformational changes involved in its catalytic cycle. These simulations revealed the detailed mechanisms of substrate binding and product release.

X-ray Crystallography: While X-ray crystallography provides static structures, it can capture different conformations by crystallizing proteins in multiple states. Such structures, called 'snapshots,' help infer dynamic behaviour.

For instance, the structural analysis of the ribosome at various stages of translation provided insights into the dynamics of ribosomal motion during protein synthesis.

Small-Angle X-ray Scattering (SAXS): SAXS is used to study protein dynamics in solution. By analysing changes in the scattering pattern of X-rays as a function of time, researchers can infer conformational changes and flexible regions within proteins.

SAXS studies on the chaperonin GroEL revealed its dynamic cycling between open and closed states, crucial for facilitating protein folding.

Hydrogen-Deuterium Exchange Mass Spectrometry (HDX-MS): HDX-MS measures the exchange of labile hydrogen atoms with deuterium in a protein's backbone amide groups. This technique provides information about the solvent accessibility and structural flexibility of proteins.

A groundbreaking HDX-MS study on protein kinase A (PKA) uncovered the dynamic assembly of its regulatory and catalytic subunits during activation.

Implications of Protein Dynamics in Function

Understanding protein dynamics has profound implications for biology and medicine. Here are a few illustrative examples:

Drug Design: Many drug targets involve proteins with dynamic active sites. Knowledge of dynamic conformations is essential for designing drugs that selectively inhibit or activate specific protein functions. The HIV-1 protease, a dynamic enzyme, exemplifies this concept, where inhibitors mimic transient conformations to block its activity.

Disease Mechanisms: Protein dynamics play a central role in the onset and progression of various diseases. For instance, in Alzheimer's disease, the dynamic behaviour of the amyloid-beta peptide leads to the formation of toxic aggregates.

Evolutionary Insights: Comparative studies of protein structures across species often reveal conserved dynamic regions. These regions are likely functionally important and can provide insights into the evolution of protein function.

Protein Engineering: Designing proteins with tailored dynamics is a promising area of research. By modulating dynamic properties, researchers can create proteins with improved catalytic efficiency or stability, leading to applications in biotechnology and medicine.

Protein dynamics are a crucial dimension of structural biology. While static structures offer valuable insights, dynamics provide the missing pieces of the puzzle, explaining how proteins perform their functions in the dynamic milieu of living cells. By combining experimental techniques and computational methods, researchers continue to unravel the intricate choreography of protein dynamics, opening new avenues for drug discovery, disease understanding, and biotechnological advancements.

15.2 Techniques for studying protein dynamics

Proteins are dynamic entities that constantly fluctuate in structure and motion to fulfil their diverse biological functions. Understanding these dynamic behaviours is crucial for gaining comprehensive insights into protein function, stability, and interactions. In this chapter, we delve into the techniques employed to study protein dynamics, ranging from classical methods to cutting-edge approaches. By deciphering the intricate dance of atoms within proteins, researchers can uncover critical details that hold the key to solving complex biological puzzles.

Classical Approaches

Nuclear Magnetic Resonance (NMR) Spectroscopy

NMR spectroscopy, renowned for its versatility in elucidating both static and dynamic protein structures, stands as a stalwart in the arsenal of techniques for studying protein dynamics. While we've previously explored its utility in determining protein structures (Chapter 3), NMR excels at unveiling dynamic processes on various timescales.

Example 1: Backbone Dynamics in Proteins

One classic application of NMR in studying protein dynamics is through the measurement of backbone amide hydrogen (N-H) exchange rates. Faster exchange rates indicate greater flexibility or conformational dynamics. Researchers have utilized this approach to investigate critical biological processes like enzymatic catalysis. For instance, a study by Boehr et al. (2006) utilized NMR to demonstrate how the dynamics of the enzyme dihydrofolate reductase (DHFR) are crucial for its catalytic activity. The enzyme's active site opens and closes dynamically, facilitating substrate binding and product release.

Example 2: Protein Folding Studies

NMR has been instrumental in unravelling the dynamics of protein folding. By examining chemical shifts, relaxation rates, and residual dipolar couplings, researchers can monitor how proteins transition between various intermediate states during folding. This knowledge has profound implications in understanding protein misfolding diseases like Alzheimer's and Parkinson's.

X-ray Crystallography

While X-ray crystallography (Chapter 2) primarily provides static snapshots of protein structures, it can also offer glimpses into conformational changes and dynamics.

Example 3: Structural Changes upon Ligand Binding

Researchers often crystallize proteins in multiple conformations, such as with and without ligands. By comparing these structures, they can discern the structural adjustments brought about by ligand binding. An iconic example is the work on the enzyme lysozyme, where the binding of a substrate was observed to induce localized structural changes in the active site. This underlines the critical role of dynamics in enzymatic catalysis.

Advanced Techniques

Molecular Dynamics (MD) Simulations

Molecular Dynamics simulations provide a computational microscope for observing protein dynamics at an atomic level. By numerically solving Newton's equations of motion, MD simulations track the positions and velocities of all atoms within a protein over time. This approach has gained immense popularity due to its ability to capture dynamics on a broad timescale, from femtoseconds to milliseconds.

Example 4: Protein Folding Pathways

MD simulations have offered unparalleled insights into the folding pathways of proteins. By simulating the folding process for a range of proteins, researchers have identified common principles governing protein folding kinetics. For example, the study of villin, an actin-binding protein, demonstrated how it folds through an intermediate state, challenging the classical two-state folding model.

Example 5: Exploring Allosteric Mechanisms

Allosteric proteins undergo conformational changes upon ligand binding at remote sites, influencing their function. MD simulations have been instrumental in deciphering these complex mechanisms. The study of allosteric regulation in the enzyme adenylate kinase elucidated how ligand binding at one site affects the dynamics of another, thus modulating enzymatic activity.

Single-Molecule Techniques

Single-molecule techniques have revolutionized the study of protein dynamics by enabling the observation of individual molecules in real-time. These methods bypass the need for ensemble averaging and provide unprecedented insights into heterogeneous dynamics.

Example 6: Single-Molecule Fluorescence Spectroscopy

Single-molecule fluorescence techniques, such as Förster resonance energy transfer (FRET), have been instrumental in exploring protein dynamics. Researchers can attach fluorescent probes to specific sites on a protein and monitor changes in fluorescence intensity or energy transfer as the protein undergoes conformational changes. This approach has been pivotal in understanding the dynamics of intrinsically disordered proteins and molecular motors like kinesin.

Example 7: Optical Tweezers

Optical tweezers use focused laser beams to trap and manipulate individual molecules. By applying controlled forces to a protein molecule, researchers can study its mechanical properties and conformational changes. This technique has been crucial in understanding the mechanical unfolding of proteins, shedding light on how proteins withstand and respond to external forces.

Hydrogen-Deuterium Exchange Mass Spectrometry (HDX-MS)

Hydrogen-deuterium exchange mass spectrometry (HDX-MS) is a powerful method for probing the solvent accessibility and flexibility of a protein's backbone amide hydrogens. This technique offers insights into the dynamics of proteins in their native solution state.

Example 8: Mapping Protein-Protein Interactions

HDX-MS can reveal changes in the solvent accessibility of amide hydrogens upon protein-protein binding. By comparing the exchange rates of specific regions, researchers can pinpoint the interaction interfaces and characterize the dynamics involved in protein-protein interactions. This has been instrumental in understanding various signalling pathways and protein complexes.

Time-Resolved Techniques

Time-resolved methods provide a dynamic dimension to structural studies by capturing protein motions on ultrafast timescales, down to femtoseconds.

Example 9: Time-Resolved X-ray Crystallography

Time-resolved X-ray crystallography employs short laser pulses to trigger conformational changes in protein crystals, which are then probed using X-ray diffraction. This technique has elucidated structural dynamics during processes such as photosynthesis, revealing how light-induced conformational changes drive energy conversion.

Example 10: Time-Resolved NMR Spectroscopy

In time-resolved NMR experiments, researchers perturb the protein system with pulses of energy and monitor the subsequent relaxation of nuclear spins. This provides information on timescales ranging from nanoseconds to seconds. Time-resolved NMR has been pivotal in studying fast protein dynamics, such as protein folding intermediates.

Future Directions

The study of protein dynamics continues to evolve with advancements in technology and methodology. Emerging techniques, such as high-speed atomic force microscopy (AFM) and ultrafast electron diffraction, promise even finer insights into the dynamic world of proteins.

Understanding protein dynamics is essential for comprehending the intricate workings of biological molecules. From classical methods like NMR and X-ray crystallography to advanced techniques like MD simulations and single-molecule experiments, researchers have a rich toolbox to explore the dynamic behaviour of proteins. These techniques, combined with ever-improving computational resources, are illuminating the fascinating choreography of atoms within the protein universe, bringing us closer to solving the mysteries of life's molecular machinery.

15.3 Implications of dynamics in function

Proteins, the workhorses of biology, are not static entities. Instead, they engage in an intricate dynamic dance that underlies their diverse functions. The concept of protein dynamics refers to the continuous motion and conformational changes that proteins undergo, from subtle fluctuations to dramatic shifts. In this subsection, we will delve into the fascinating world of protein dynamics and explore how these movements are intimately connected to the functions they perform. Through a combination of experimental evidence and computational simulations, we will uncover the profound implications of dynamics in protein function.

Understanding Protein Dynamics

Imagine a protein as a complex three-dimensional structure, with a multitude of atoms interconnected by covalent bonds. At first glance, it may seem like a static sculpture, but when we zoom in, we discover a vibrant molecular world. Proteins are in a constant state of motion, resembling a bustling city where atoms jostle and vibrate. These movements occur on various timescales, from picoseconds to seconds, and encompass a wide range of motions, such as bond stretching, bending, and torsional rotations.

One of the fundamental aspects of protein dynamics is thermal fluctuations. At room temperature, proteins are not frozen in place but exist in a dynamic equilibrium of multiple conformations. These fluctuations are essential for their function, as they enable proteins to explore different structural states and interact with other molecules.

Case Study 1: Enzymatic Catalysis

To appreciate the implications of dynamics in protein function, let's turn our attention to enzymatic catalysis, one of the most crucial functions in biology. Enzymes are proteins that facilitate chemical reactions by lowering the activation energy barrier. The traditional "lock-and-key" model of enzyme-substrate interaction has evolved into a more dynamic and nuanced perspective.

Enzymes exhibit dynamic flexibility in their active sites, where substrates bind and reactions occur. Molecular dynamics simulations and nuclear magnetic resonance (NMR) experiments have revealed that active site residues can undergo conformational changes upon substrate binding. This dynamic behaviour allows enzymes to adapt their active sites to fit substrates of various sizes and shapes.

For example, in the enzyme dihydrofolate reductase (DHFR), which plays a critical role in DNA synthesis, the active site undergoes a conformational change upon substrate binding, closing around the bound substrate. This conformational change not only stabilizes the transition state of the reaction but also protects the reactive intermediates from the surrounding solvent, enhancing the catalytic efficiency.

Case Study 2: Allosteric Regulation

Another captivating aspect of protein dynamics is allosteric regulation, where a molecule binding at one site on a protein induces a structural change at a distant site, thereby modulating the protein's function. Allosteric proteins act as molecular switches, and their dynamic nature is central to their regulatory function.

One classic example is haemoglobin, the protein responsible for oxygen transport in red blood cells. Haemoglobin exists in two distinct conformations: the relaxed (R) state, which has a high affinity for oxygen, and the tense (T) state, with a lower affinity for oxygen. When oxygen binds to one subunit of haemoglobin, it induces a conformational change that propagates through the protein, shifting it from the T state to the R state. This allosteric transition enhances the protein's oxygen-carrying capacity, enabling efficient oxygen delivery to tissues.

Computational Tools Unveiling Dynamics

While experimental techniques like NMR and X-ray crystallography provide critical insights into protein dynamics, computational methods have become indispensable tools in this field. Molecular dynamics simulations, in particular, have revolutionized our understanding of protein dynamics by allowing us to track the movements of thousands of atoms over time.

Through molecular dynamics simulations, researchers can observe how proteins flex, twist, and breathe. These simulations provide a dynamic "movie" of protein behaviour, revealing the intricate interplay between structure and motion. For example, researchers have used simulations to study the dynamic behaviour of G protein-coupled receptors (GPCRs), a family of proteins involved in cell signalling. GPCRs undergo complex conformational changes upon ligand binding, which are challenging to capture experimentally. Molecular dynamics simulations have provided valuable insights into these dynamic processes, shedding light on receptor activation and downstream signalling events.

Biological Significance of Dynamics

The dynamic nature of proteins has profound biological implications. Consider that the human body contains trillions of cells, each housing thousands of different proteins, all engaged in dynamic interactions. These interactions are fundamental to processes such as signal transduction, muscle contraction, and immune responses.

Case Study 3: Immune System Recognition

One of the most remarkable examples of protein dynamics in action is the recognition of antigens by antibodies in the immune system. Antibodies are Y-shaped proteins that can bind to a vast array of antigens, including pathogens like viruses and bacteria. The antigen-binding site of an antibody is highly flexible, allowing it to adapt to the unique shape and chemical composition of different antigens.

As an antibody encounters an antigen, its dynamic paratope (antigen-binding site) undergoes conformational changes to achieve a complementary fit. This binding event triggers a cascade of immune responses, leading to the destruction or neutralization of the invading pathogen. Without the dynamic flexibility of antibodies, our immune system would be far less effective at recognizing and combating diverse pathogens.

Dynamic Complexity in Drug Design

Understanding protein dynamics has become increasingly important in drug discovery. Many drug targets, such as enzymes and receptors, undergo dynamic changes upon ligand binding. Researchers now recognize that designing drugs solely based on static protein structures may lead to suboptimal results. Instead, drug developers are turning to dynamic structural information to design more effective therapeutics.

Case Study 4: HIV Protease Inhibition

The enzyme HIV protease is a critical drug target in the treatment of HIV/AIDS. Early drug design efforts focused on static crystal structures of the protease, leading to the development of protease inhibitors. However, these inhibitors faced challenges due to the protease's dynamic nature.

Researchers turned to molecular dynamics simulations to gain insights into the dynamic behaviour of the protease and its interactions with inhibitors. These simulations revealed that the protease undergoes significant structural fluctuations, even in the presence of inhibitors. By considering these dynamics, researchers were able to design more potent protease inhibitors that could better adapt to the dynamic nature of the target, leading to improved HIV/AIDS therapies.

Protein dynamics are not mere background noise in the world of structural biology; they are the orchestration of life's molecular symphony. From enzymatic catalysis to immune recognition and drug design, the dynamic behaviour of proteins underpins their remarkable functionality. As our tools for studying protein dynamics continue to advance, we can expect to uncover even more intricate details of this dynamic dance and harness this knowledge for novel therapeutic interventions and a deeper understanding of life at the molecular level. The dynamic complexity of proteins remains a testament to the elegance of nature's design.

Chapter 16: Structural Biology in Drug Discovery

16.1 Role of high-resolution structures in drug development

In an ever-changing landscape of pharmaceutical research and development, the role of high-resolution protein structures is pivotal. These structures serve as invaluable blueprints that guide the rational design of new drugs and therapeutics. In this chapter, we delve into the multifaceted ways in which high-resolution structures contribute to drug development, showcasing real-world examples and highlighting the significant impact they have had on the industry.

Target Identification and Validation

The journey of drug development often begins with the identification and validation of suitable drug targets. High-resolution protein structures play a crucial role in this initial phase by providing insights into the three-dimensional (3D) architecture of target proteins. Understanding the structure of a target allows researchers to identify key binding sites, predict potential ligand interactions, and assess the druggability of the target.

Example 1: The Case of Gleevec (Imatinib)

A remarkable example of target identification and validation is the development of Gleevec, a breakthrough drug for chronic myeloid leukaemia (CML). Researchers used the crystal structure of the Abelson kinase domain (Abl kinase), a target in CML, to design Gleevec. The structure revealed a unique pocket that could be targeted by a small molecule, leading to the development of this highly effective drug.

Structure-Based Drug Design (SBDD)

Structure-based drug design (SBDD) is a powerful approach that leverages high-resolution protein structures to design novel drug candidates with high specificity and affinity. By visualizing the target in atomic detail, researchers can computationally screen a vast library of compounds and predict which ones are most likely to bind effectively.

Example 2: Design of HIV Protease Inhibitors

The development of HIV protease inhibitors provides a classic example of SBDD. The crystal structure of the HIV protease enzyme revealed its active site's precise geometry. Researchers used this structural information to design inhibitors like ritonavir and darunavir, which fit snugly into the active site and block the enzyme's function, thus inhibiting viral replication.

Hit Identification and Lead Optimization

High-resolution structures not only aid in the initial design of drug candidates but also play a pivotal role in hit identification and lead optimization phases. Researchers can determine how potential drug candidates interact with the target at the atomic level, allowing for fine-tuning of their chemical properties to enhance binding affinity and specificity.

Example 3: Optimizing Kinase Inhibitors

Kinase inhibitors, used in the treatment of various cancers, have undergone extensive lead optimization. By studying the crystal structures of kinase-inhibitor complexes, researchers iteratively improved the drug candidates' structures, enhancing their potency and minimizing off-target effects.

Understanding Resistance Mechanisms

As drug resistance remains a significant challenge in medicine, high-resolution structures help researchers understand the molecular basis of resistance. By comparing the structures of drug-resistant and sensitive target proteins, scientists can uncover mutations or conformational changes that confer resistance.

Example 4: Resistance to Antibiotics

The rise of antibiotic resistance is a global health concern. High-resolution structures of bacterial enzymes, such as beta-lactamases, have revealed the structural basis of resistance to antibiotics like penicillin. This knowledge informs the development of new antibiotics that can bypass resistance mechanisms.

Predicting Drug-Drug Interactions

In polypharmacy scenarios, where patients take multiple medications simultaneously, predicting potential drug-drug interactions is crucial for patient safety. High-resolution structures can help identify potential binding sites for multiple drugs on a target protein, facilitating predictions of interaction likelihood.

Example 5: Drug Interactions with Cytochrome P450 Enzymes

Cytochrome P450 enzymes play a central role in drug metabolism. Researchers use the crystal structures of these enzymes to predict how different drugs may compete for binding sites or inhibit each other's metabolism, influencing drug interactions and dosing regimens.

Reducing Adverse Effects

High-resolution structures enable researchers to design drugs with a higher degree of specificity, reducing the risk of off-target effects. By precisely understanding the target's structure, drug developers can avoid unintended interactions with related proteins.

Example 6: Selective Serotonin Reuptake Inhibitors (SSRIs)

Selective serotonin reuptake inhibitors (SSRIs), used to treat depression, are designed to selectively inhibit the serotonin transporter protein. High-resolution structures of this transporter have aided in the development of SSRIs with improved selectivity, reducing side effects associated with off-target interactions.

Accelerating Drug Development

The integration of structural biology into drug development pipelines has accelerated the process of bringing new therapeutics to market. High-resolution structures streamline the optimization of drug candidates, reducing the time and resources required for preclinical and clinical development.

Example 7: COVID-19 Vaccine Development

The rapid development of COVID-19 vaccines is a testament to the impact of structural biology. The determination of the SARS-CoV-2 spike protein's structure allowed for the design of vaccines that target this protein, accelerating vaccine development and deployment during the pandemic.

Personalized Medicine

High-resolution structures have also paved the way for personalized medicine approaches. By considering an individual's genetic variations and the 3D structure of their target proteins, it's possible to tailor drug treatments to maximize efficacy and minimize side effects.

Example 8: Cystic Fibrosis Transmembrane Conductance Regulator (CFTR) Modulators

In cystic fibrosis, drugs like ivacaftor and lumacaftor target specific CFTR mutations. High-resolution structures of CFTR variants guide the design of modulators that can restore normal protein function based on the patient's genotype.

Beyond Small Molecules: Biologics and Antibodies

While high-resolution structures are traditionally associated with small molecule drug development, they also play a significant role in the design of biologics, including monoclonal antibodies and protein-based therapeutics.

Example 9: Monoclonal Antibodies for Cancer Immunotherapy

Monoclonal antibodies like trastuzumab (Herceptin) for breast cancer and pembrolizumab (Keytruda) for immunotherapy are designed based on the 3D structures of their target proteins, providing high specificity and minimal side effects.

High-resolution protein structures have become indispensable tools in drug development. They guide target identification, enable rational drug design, facilitate lead optimization, and offer insights into resistance mechanisms and drug interactions. By accelerating drug development and improving therapeutic specificity, high-resolution structures continue to transform the pharmaceutical industry, offering new hope for the treatment of various diseases and conditions. As structural biology techniques advance, their impact on drug development is poised to grow even further, ushering in an era of more effective and personalized medicines.

16.2 Case studies of successful drug design based on protein structures

In the ever-evolving landscape of drug discovery, the integration of structural biology has emerged as a powerful tool. By elucidating the three-dimensional structures of proteins and their complexes with ligands, researchers gain profound insights into the molecular mechanisms of diseases. This knowledge forms the foundation for rational drug design, enabling the development of more effective and targeted therapeutics. In this chapter, we will explore several remarkable case studies that exemplify the successful application of protein structures in drug design.

Imatinib: A Paradigm Shift in Cancer Treatment

One of the most celebrated success stories in drug design is the development of Imatinib (trade name Gleevec or Glivec) for the treatment of chronic myeloid leukaemia (CML) and gastrointestinal stromal tumours (GISTs). Prior to Imatinib, CML was a grim diagnosis with limited treatment options. This small molecule inhibitor revolutionized the field by specifically targeting the tyrosine kinase activity of the BCR-ABL fusion protein, which is the hallmark of CML.

Protein Structure Insights: The breakthrough in Imatinib's development came from the determination of the crystal structure of the Abl kinase domain in complex with Imatinib. This structure revealed how the drug binds to the active site of the kinase, locking it in an inactive conformation. Imatinib's selectivity for BCR-ABL over other kinases was also attributed to the specific interactions observed in the crystal structure.

Outcome: Imatinib's approval by the FDA in 2001 marked a turning point in cancer treatment. It has since become a standard of care for CML patients, transforming the prognosis from a life-threatening condition to a manageable chronic disease.

Tamiflu: Battling the Influenza Virus

Influenza, with its seasonal outbreaks and potential for pandemics, poses a significant global health threat. Tamiflu (oseltamivir) is an antiviral drug designed to combat influenza by inhibiting the neuraminidase enzyme of the virus, preventing its spread within the host.

Protein Structure Insights: The foundation for Tamiflu's design was the crystal structure of the influenza neuraminidase. This structure revealed the active site pocket where Tamiflu binds, preventing the virus from cleaving sialic acid residues and releasing progeny viruses from infected cells.

Outcome: Tamiflu has been instrumental in managing influenza outbreaks and reducing the severity and duration of symptoms. Its structural insights have also paved the way for the development of other neuraminidase inhibitors.

HIV Protease Inhibitors: Tackling the AIDS Pandemic

The human immunodeficiency virus (HIV) has claimed millions of lives worldwide, leading to the AIDS pandemic. A crucial breakthrough in the fight against HIV came with the development of protease inhibitors, which block the activity of the viral protease enzyme, essential for viral maturation.

Protein Structure Insights: The design of HIV protease inhibitors was guided by the crystallographic structures of the viral protease in complex with various inhibitors. These structures revealed the precise binding interactions and conformational changes required for effective inhibition.

Outcome: Protease inhibitors like ritonavir, saquinavir, and lopinavir have significantly improved the prognosis for HIV-positive individuals. They have been instrumental in reducing viral loads and slowing disease progression.

Herceptin: Targeted Therapy for Breast Cancer

Breast cancer is a heterogeneous disease, and the development of targeted therapies has been a game-changer. Herceptin (trastuzumab) is a monoclonal antibody designed to specifically target the human epidermal growth factor receptor 2 (HER2), which is overexpressed in a subset of breast cancers.

Protein Structure Insights: Understanding the structure of HER2 and its interaction with Herceptin allowed for the rational design of the antibody. The crystal structure of the HER2 extracellular domain in complex with Herceptin revealed how the antibody binds to HER2, inhibiting its signalling pathways.

Outcome: Herceptin has become a standard treatment for HER2-positive breast cancer patients, significantly improving survival rates and quality of life.

Pembrolizumab and Nivolumab: Unleashing the Immune System against Cancer

Immunotherapy has revolutionized cancer treatment by harnessing the body's immune system to target cancer cells. Pembrolizumab (Keytruda) and nivolumab (Opdivo) are immune checkpoint inhibitors that block the PD-1 receptor on T cells, allowing them to mount a more effective anti-tumour response.

Protein Structure Insights: The success of these inhibitors can be attributed to the detailed structural knowledge of the PD-1 receptor and its interaction with ligands. Crystallographic studies elucidated the binding interface and the role of these inhibitors in preventing the immune suppression induced by cancer cells.

Outcome: Pembrolizumab and nivolumab have demonstrated remarkable efficacy across various cancer types, leading to durable responses and improved long-term survival in patients who had exhausted other treatment options.

Monoclonal Antibodies against SARS-CoV-2: A Rapid Response to a Global Pandemic

The COVID-19 pandemic brought the world to a standstill, underscoring the urgent need for effective therapeutics. Monoclonal antibodies such as Regeneron's casirivimab and imdevimab, and Eli Lilly's bamlanivimab and etesevimab, were developed as passive immunization agents to neutralize the SARS-CoV-2 virus.

Protein Structure Insights: These monoclonal antibodies were designed based on the structural characterization of the SARS-CoV-2 spike protein, particularly its receptor-binding domain (RBD). Crystal structures of the RBD in complex with these antibodies demonstrated the precise binding epitopes and neutralizing mechanisms.

Outcome: Monoclonal antibody therapies have played a crucial role in reducing the severity of COVID-19 and preventing hospitalizations. They have become an essential tool in the battle against the pandemic.

In each of these case studies, the critical role of high-resolution protein structures in drug design is evident. From small molecule inhibitors to monoclonal antibodies, the detailed knowledge of the target proteins and their interactions with ligands has paved the way for the development of therapeutics that have transformed the landscape of disease management. These successes not only highlight the power of structural biology in drug discovery but also underscore its potential in addressing the most challenging health crises of our time. As we continue to advance our understanding of protein structures and their roles in diseases, the prospects for innovative drug design and personalized medicine only grow brighter.

Chapter 17: Structural Genomics and Proteomics

17.1 Large-scale protein structure determination initiatives

While signifying structural biology, the pursuit of understanding the three-dimensional structures of proteins has led to groundbreaking initiatives on a grand scale. These initiatives, often referred to as large-scale protein structure determination initiatives, are monumental endeavours that seek to map the structures of thousands of proteins to unravel the mysteries of life. In this section, we will explore some of the most prominent large-scale initiatives, their objectives, methodologies, and the profound impact they have had on advancing our understanding of biology.

The Protein Data Bank (PDB): A Cornerstone of Structural Biology

At the heart of structural biology lies the Protein Data Bank (PDB), a pivotal resource that catalogues experimentally determined protein structures. Established in 1971, the PDB has been instrumental in shaping the field of structural biology. It serves as a repository for atomic coordinates and structural information, allowing researchers worldwide to access and analyse protein structures. The PDB has grown exponentially over the years, both in the number of deposited structures and in its impact on scientific research.

The PDB contained over 180,000 structures, covering a vast spectrum of biological molecules, from proteins and nucleic acids to small molecules and macromolecular complexes. This wealth of structural data has fuelled discoveries in fields ranging from drug design to evolutionary biology.

The Structural Genomics Initiative

One of the pioneering large-scale initiatives in structural biology is the Structural Genomics (SG) Initiative. Launched in the late 1990s, the SG Initiative aimed to systematically determine the three-dimensional structures of proteins on a genomic scale. The primary objective was to bridge the gap between the increasing number of protein sequences emerging from genome sequencing projects and the limited availability of experimental structures.

The SG Initiative employed a high-throughput approach, utilizing a combination of X-ray crystallography, NMR spectroscopy, and other techniques to rapidly determine protein structures. It often focused on proteins with unknown functions, providing invaluable insights into uncharted regions of the protein universe.

Notable Successes of the Structural Genomics Initiative

One of the most significant accomplishments of the SG Initiative was the determination of structures for numerous hypothetical and conserved proteins. For example, the Northeast Structural Genomics Consortium (NESG) played a pivotal role in solving the structure of a protein known as YciH from Escherichia coli. Despite having an unknown function, the protein's structure revealed striking similarities to a subunit of the bacterial ribosome, shedding light on its potential role in protein synthesis.

Another remarkable achievement of the SG Initiative was the structural characterization of proteins from pathogenic organisms. The Seattle Structural Genomics Centre for Infectious Disease (SSGCID), for instance, focused on proteins from infectious agents like the Ebola virus and the bacterium responsible for tuberculosis. These efforts contributed to our understanding of the molecular mechanisms underlying infectious diseases and offered targets for drug development.

The Structural Biology and Structural Proteomics Initiative

Parallel to the SG Initiative, the Structural Biology and Structural Proteomics (SBSP) Initiative emerged as a global effort to decipher protein structures. Led by organizations like the National Institute of General Medical Sciences (NIGMS) in the United States, the SBSP Initiative aimed to accelerate the pace of structural biology research.

One of the key components of the SBSP Initiative was the establishment of specialized research centres, each with a particular focus on specific aspects of structural biology. These centres combined resources and expertise to tackle a wide array of biological questions.

Impact on Drug Discovery

Large-scale structural initiatives have had a profound impact on drug discovery. The structural information generated has been instrumental in rational drug design, enabling scientists to develop new therapeutics with greater precision and efficiency. For instance, the structures of protein targets from disease-causing organisms, such as those determined by SSGCID, have paved the way for the design of novel antibiotics and antiviral drugs.

Additionally, the SBSP Initiative has significantly contributed to our understanding of human proteins and their functions. This knowledge has implications for drug development in areas ranging from cancer to neurodegenerative diseases. With structural insights into drug targets, pharmaceutical companies can streamline drug development pipelines and increase the success rate of new drug candidates.

The Role of Automation and Robotics

Automation and robotics have been instrumental in the success of large-scale protein structure determination initiatives. High-throughput platforms equipped with robotic systems have revolutionized the process of protein expression, purification, and crystallization. These advancements have not only accelerated structure determination but have also improved the quality and reproducibility of results.

Moreover, automated data collection at synchrotron facilities has become standard practice in X-ray crystallography. This has enabled researchers to rapidly collect diffraction data from a large number of crystals, further enhancing the efficiency of structural studies.

Challenges and Future Directions

Despite the remarkable achievements of large-scale initiatives, several challenges persist. The structural characterization of membrane proteins, which play crucial roles in biology and disease, remains a formidable task. Moreover, tackling the dynamic and heterogeneous nature of certain protein complexes requires innovative approaches in cryo-EM and NMR spectroscopy.

Looking ahead, future initiatives may explore the integration of structural data with other omics data, such as genomics and proteomics, to gain a more comprehensive understanding of cellular processes. Furthermore, advances in artificial intelligence and machine learning promise to expedite data analysis and enhance our ability to predict protein structures accurately.

Large-scale protein structure determination initiatives have revolutionized the field of structural biology. They have expanded our knowledge of protein structures, enabled drug discovery, and provided valuable insights into biological processes. As technology continues to advance and new challenges emerge, these initiatives will play a pivotal role in unlocking the secrets of life at the molecular level.

17.2 Advances in high-throughput structure determination

In the land of structural biology, the pursuit for high-resolution protein structures has traditionally been a laborious and time-consuming endeavour. However, in recent years, significant strides have been made in the development of high-throughput methods that promise to revolutionize the field. High-throughput structure determination holds the potential to accelerate our understanding of biological macromolecules and their functions, making it an exciting frontier in structural biology.

The High-Throughput Paradigm

High-throughput structure determination can be likened to a high-speed assembly line, where numerous proteins can be rapidly processed and their structures elucidated. Traditionally, determining the three-dimensional structure of a single protein could take months or even years. High-throughput approaches aim to reduce this timeframe drastically, potentially enabling the structural analysis of hundreds or thousands of proteins in a shorter span. This paradigm shift is crucial for addressing the structural biology bottleneck and facilitating research in various fields, including drug discovery, functional genomics, and systems biology.

Structural Genomics Initiatives

One of the most significant drivers of high-throughput structure determination is the establishment of structural genomics initiatives. These large-scale, collaborative efforts aim to systematically determine the structures of proteins on a genome-wide scale. An early milestone in this endeavour was the Structural Genomics Initiative (SGI), launched in the early 2000s.

The SGI focused on the structural characterization of proteins from various organisms, including bacteria, archaea, and eukaryotes. To achieve this ambitious goal, SGI employed a combination of X-ray crystallography and NMR spectroscopy, complemented by advanced automation and robotics. By streamlining and standardizing the structural determination process, SGI significantly increased throughput.

Example: SGI Success Story

One notable success story from the SGI was the structural elucidation of a large number of bacterial proteins, many of which were of unknown function. These structures provided valuable insights into bacterial biology and opened new avenues for antibiotic development. For instance, the crystal structure of a previously uncharacterized bacterial enzyme revealed a unique catalytic mechanism, inspiring the design of novel antibiotics that targeted this enzyme.

Automation and Robotics

Automation plays a pivotal role in high-throughput structure determination. It involves the use of advanced robotics to handle the numerous steps involved in sample preparation, data collection, and data analysis. The benefits of automation are manifold: it reduces human error, increases the speed of experiments, and allows for the parallel processing of multiple samples.

Example: Automated Crystallization

In X-ray crystallography, automating the crystallization process has been a game-changer. Robots can set up thousands of crystallization trials in a short period, testing various conditions to optimize crystal growth. This approach led to a significant increase in the success rate of crystallization experiments and a higher throughput of protein structures.

Miniaturization and Microfluidics

Advancements in miniaturization and microfluidic technologies have further propelled high-throughput structure determination. Miniaturization allows researchers to work with smaller volumes of reagents, reducing costs and increasing efficiency.

Microfluidics, on the other hand, enables precise manipulation of tiny liquid droplets, facilitating high-throughput screening experiments.

Example: Microfluidic NMR

In NMR spectroscopy, microfluidic devices can rapidly mix protein samples with different ligands or conditions, allowing for high-throughput screening of protein-ligand interactions. This approach not only accelerates data collection but also conserves precious protein samples, which is particularly important for challenging targets.

Data Analysis Pipelines

High-throughput structure determination generates vast amounts of data that must be efficiently processed and analysed. To meet this challenge, dedicated data analysis pipelines have been developed. These pipelines employ algorithms and software tools to automate data processing steps, including data reduction, phasing, and model building.

Example: AutoDep Input Pipeline

AutoDep Input is an automated pipeline used in X-ray crystallography that streamlines the process of model building. It utilizes algorithms to interpret electron density maps, automatically placing atoms in the correct positions. This significantly accelerates the time required to generate an initial protein structure.

The Impact on Drug Discovery

High-throughput structure determination has profound implications for drug discovery and development. Understanding the three-dimensional structure of a drug target protein is essential for rational drug design. Traditional approaches to structure-based drug discovery often faced limitations due to the slow pace of structural characterization. High-throughput methods have addressed these limitations and are transforming the field.

Example: Rapid Structure-Based Drug Design

Imagine a pharmaceutical company aiming to develop a new cancer drug targeting a specific protein. In the past, obtaining the high-resolution structure of this protein might have taken years. With high-throughput methods, the structure can be determined in a matter of weeks. This rapid access to structural information allows medicinal chemists to design drug candidates with high precision, leading to accelerated drug development timelines.

Expanding the Structural Universe

High-throughput structure determination has also expanded the structural universe by enabling the study of challenging protein classes that were previously considered "undruggable" or too complex to handle. Membrane proteins, for instance, are notoriously difficult to crystallize and study, yet they are essential drug targets. High-throughput methods have made significant strides in membrane protein structural biology.

Example: Membrane Protein Structures

Researchers using high-throughput approaches have successfully determined the structures of numerous membrane proteins, shedding light on their functions and paving the way for the development of drugs targeting these proteins. This includes G-protein coupled receptors (GPCRs), which are key drug targets in various therapeutic areas.

Bridging the Gap with Cryo-Electron Microscopy

Cryo-electron microscopy (cryo-EM) is another technique that has seen remarkable advancements in high-throughput structural determination. While traditionally considered a method for larger complexes, cryo-EM is now being applied to smaller protein targets, blurring the lines between traditional structural biology techniques.

Example: Single-Particle Cryo-EM

Single-particle cryo-EM has become a high-throughput method for determining the structures of diverse macromolecules, including proteins and protein complexes. Automation and the use of advanced detectors have made it possible to rapidly collect high-quality data, even for challenging samples. This approach has democratized cryo-EM, making it accessible to a broader scientific community.

Challenges and Future Directions

While high-throughput structure determination holds immense promise, it is not without its challenges. One significant challenge is the need for rigorous validation of high-throughput structures. As the pace of structural determination accelerates, it becomes crucial to maintain the highest standards of data quality and accuracy.

Moreover, high-throughput methods are not universally applicable. Some proteins remain recalcitrant to crystallization or NMR analysis, and certain complexes may resist structural determination. Therefore, complementary approaches, such as computational modelling and hybrid methods, will continue to play a vital role in structural biology.

The future of high-throughput structure determination holds exciting prospects. Integration with artificial intelligence and machine learning is expected to further automate and accelerate data analysis. Additionally, the development of novel techniques, such as serial crystallography and time-resolved structural biology, will push the boundaries of what is possible in the quest for high-resolution protein structures.

The field of structural biology is undergoing a rapid transformation thanks to advances in high-throughput structure determination. Structural genomics initiatives, automation, miniaturization, microfluidics, and cryo-EM are all contributing to the acceleration of structural biology research. These advancements not only impact drug discovery but also expand our understanding of the structural universe, paving the way for breakthroughs in diverse scientific disciplines. As the pace of high-throughput structural determination continues to quicken, the future promises a deeper and more comprehensive view of the molecular world.

17.3 Functional insights from structural genomics

Structural genomics, a subfield of genomics, has made remarkable strides in the last few decades, revolutionizing our understanding of biology at the molecular level. It focuses on the systematic determination of the three-dimensional structures of proteins on a genome-wide scale. While the primary goal of structural genomics is to provide a comprehensive structural map of the proteome, it goes far beyond merely cataloguing structures. The wealth of data generated through structural genomics initiatives has opened up exciting avenues for deciphering the functional aspects of proteins, enabling us to unravel their roles in health, disease, and evolution. In this chapter, we will delve into the world of functional insights obtained from structural genomics, exploring how these structures have shed light on protein function, drug discovery, and beyond.

Understanding Protein Function through Structure

A fundamental axiom in molecular biology is that "structure determines function." This principle is at the core of structural genomics. As we elucidate the three-dimensional architecture of proteins, we gain a more profound understanding of their function and mechanisms. Let's explore how this plays out in practice.

Annotating Proteins of Unknown Function

A significant achievement of structural genomics has been the annotation of proteins with previously unknown functions. In many genomes, a substantial fraction of genes encode proteins whose roles remain enigmatic. The structural information can offer crucial clues. For instance, if a protein's structure resembles that of a known enzyme, it might suggest a catalytic function. Conversely, a structural similarity to a protein with a well-established role in DNA binding can hint at a DNA-binding function. The Centre for Structural Genomics of Infectious Diseases (CSGID), for example, has been instrumental in characterizing proteins from pathogenic microorganisms. By determining the structures of these proteins, researchers have uncovered potential drug targets and mechanisms of pathogenicity.

Example: Mycobacterium tuberculosis Pantothenate Synthetase

Pantothenate synthetase is an essential enzyme involved in Coenzyme A biosynthesis in Mycobacterium tuberculosis. The structure of this protein revealed its active site and the binding pocket for its substrate. This information was pivotal in the design of inhibitors targeting this enzyme, which holds promise as a drug target against tuberculosis.

Elucidating Catalytic Mechanisms

Structural genomics has played a pivotal role in elucidating the catalytic mechanisms of various enzymes. By visualizing the active sites and substrate binding regions in three dimensions, researchers gain insights into how these enzymes facilitate specific chemical reactions. Such knowledge is invaluable for biotechnology, as it enables the rational design of enzymes with enhanced catalytic properties.

Example: DNA Polymerase

The high-resolution structures of DNA polymerases have revealed the intricate details of DNA replication. They showcase how these enzymes select and incorporate the correct nucleotides during DNA synthesis and how they proofread to correct errors. Understanding these mechanisms has implications not only for basic biology but also for cancer research and drug development.

Investigating Protein-Protein Interactions

Proteins rarely work in isolation; instead, they often form complexes with other proteins to carry out their functions. Structural genomics has facilitated the study of these protein-protein interactions by providing detailed structures of protein complexes. This information is crucial for deciphering signalling pathways, understanding disease mechanisms, and designing drugs that disrupt harmful interactions.

Example: G Protein-Coupled Receptor (GPCR) Complexes

GPCRs are a class of proteins involved in cellular signalling. Structural genomics efforts have yielded insights into the structures of GPCRs in complex with their ligands, showing how these receptors interact with signalling molecules. This knowledge has implications for drug discovery, as many pharmaceuticals target GPCRs.

Drug Discovery and Target Identification

The pharmaceutical industry has reaped significant benefits from structural genomics. Protein structures have become invaluable tools for drug discovery and development. Understanding the three-dimensional structure of a target protein can guide the design of small molecules that specifically bind to it, modulating its activity and, in many cases, inhibiting disease progression.

Rational Drug Design

With the structural information about a target protein, medicinal chemists can engage in rational drug design. This approach involves using the protein's structure to design small molecules or peptides that fit precisely into the active site, blocking or enhancing its activity as needed. This precision reduces the likelihood of off-target effects and increases the chances of drug success.

Example: HIV Protease

The HIV protease is a critical enzyme for viral replication. Its structure was a prime target for structural genomics, leading to the design of protease inhibitors like lopinavir and ritonavir. These drugs have significantly improved the prognosis for individuals living with HIV.

Structure-Guided Drug Optimization

Beyond initial drug design, structural genomics has played a role in optimizing drug candidates. Researchers use structural data to understand how drugs interact with their targets and to improve drug binding affinity and specificity.

Example: Antibiotics

Structural studies of bacterial ribosomes have informed the development of antibiotics like linezolid and tetracycline. By visualizing how these antibiotics bind to the ribosome, researchers have refined their structures to increase efficacy and reduce resistance.

Evolutionary Insights

Protein structures offer a window into evolutionary history. By comparing the structures of related proteins across species, researchers can trace the evolutionary changes that have occurred over millions of years. This comparative structural genomics approach provides insights into the origins and adaptations of proteins.

Example: Haemoglobin Evolution

Comparative structural studies of haemoglobin, the oxygen-carrying protein in blood, have revealed how this protein has evolved in different species to adapt to their specific physiological needs. The structure of foetal haemoglobin, for instance, differs from that of adult haemoglobin, reflecting adaptations for oxygen transport in developing organisms.

Beyond Proteins: RNA and Complexes

While the focus of structural genomics has predominantly been on proteins, it has also expanded to include RNA molecules and protein-RNA complexes. These efforts have illuminated the structures of ribosomes, ribozymes, and RNA-binding proteins, contributing to our understanding of gene regulation, RNA processing, and disease mechanisms.

Example: Ribosome Structure

Determining the structures of ribosomes from various species has unveiled their intricate architecture and provided insights into the mechanisms of protein synthesis. This knowledge has implications not only for fundamental biology but also for the development of antibiotics that target bacterial ribosomes.

Structural genomics has evolved from a mere structural cataloguing endeavour to a powerful tool for uncovering the functional mysteries of biological macromolecules. Through a myriad of examples, we have seen how high-resolution protein structures have illuminated protein function, enabled drug discovery, provided insights into evolution, and expanded our understanding of the molecular intricacies of life. As structural genomics continues to advance, its impact on biology and medicine is bound to grow, offering new perspectives on the complex web of molecular interactions that underpin cellular life.

Chapter 18: Emerging Technologies in Protein Structure Determination

18.1 New and upcoming techniques in structural biology

Technological advancements, driven by a search for higher resolution and more efficient methods, have paved the way for innovative techniques that promise to revolutionize our understanding of protein structures and their functional implications. In this subsection, we delve into some of the most exciting emerging techniques in structural biology, providing a glimpse into the future of this field.

Cryo-Electron Tomography (Cryo-ET)

Cryo-electron tomography (Cryo-ET) is a groundbreaking technique that extends the capabilities of traditional single-particle cryo-electron microscopy (cryo-EM). While single-particle cryo-EM is well-suited for high-resolution imaging of isolated macromolecular complexes, Cryo-ET allows for the 3D reconstruction of cellular structures in their native context. Researchers can capture a series of 2D images of a target structure from different angles and use computational methods to generate a 3D reconstruction.

This technique opens up new avenues for studying dynamic processes within cells, such as intracellular transport, virus assembly, and cellular organelles. For example, researchers have used Cryo-ET to visualize the Zika virus in 3D within infected host cells, providing insights into the viral life cycle and potential drug targets.

In-cell NMR Spectroscopy

Traditional NMR spectroscopy relies on purified samples in solution, limiting its applicability to proteins that can be isolated and purified. In-cell NMR spectroscopy, however, overcomes this limitation by allowing researchers to study proteins directly within living cells. By introducing isotopically labelled molecules into cells and recording NMR spectra, scientists can monitor protein structures and dynamics in their native environment.

This technique has immense potential for investigating intracellular protein-protein interactions, folding processes, and conformational changes in response to cellular cues. It has been used to study the structural changes of proteins involved in neurodegenerative diseases, shedding light on the molecular mechanisms underlying conditions like Alzheimer's and Parkinson's disease.

Serial Crystallography with X-ray Free-Electron Lasers (XFELs)

X-ray crystallography has been a stalwart in structural biology for decades, but it faces limitations when dealing with radiation-sensitive samples or extremely small crystals. Enter X-ray Free-Electron Lasers (XFELs), which produce ultra-bright and extremely short X-ray pulses, making them ideal for studying such challenging samples.

In serial crystallography, thousands of tiny crystals are exposed to XFEL pulses, and data from each crystal are merged to generate a complete diffraction dataset. This approach has been used to determine high-resolution structures of membrane proteins and photosynthetic complexes that were previously intractable. XFELs are expected to become more accessible, leading to a broader range of applications.

Cryo-Electron Microscopy of Radiation-Sensitive Samples

One of the limitations of cryo-EM has been its sensitivity to radiation, which can damage biological samples during imaging. To address this issue, emerging techniques like "MicroED" (Micro-Electron Diffraction) and "Time-resolved Cryo-EM" are making waves in the field.

MicroED is a cryo-EM technique that uses extremely small 3D protein crystals and electron diffraction data to determine atomic-level structures. It has been used to solve the structures of challenging targets like amyloid fibrils implicated in neurodegenerative diseases. Time-resolved Cryo-EM, on the other hand, enables the visualization of dynamic processes by rapidly freezing samples at different time points. This promises to elucidate structural changes in proteins during biochemical reactions.

Protein Structure Determination by Solid-State NMR

While solution NMR has been widely used for studying protein structures, solid-state NMR is gaining prominence for its ability to investigate proteins in non-crystalline states, such as membrane proteins embedded in lipid bilayers or fibrillar aggregates. Solid-state NMR can provide crucial insights into the structure and dynamics of proteins under physiological conditions.

This technique has been instrumental in characterizing the structures of amyloid fibrils associated with diseases like Alzheimer's and Parkinson's, paving the way for targeted drug design. It is expected that advances in solid-state NMR technology will make it more accessible and applicable to a broader range of biological systems.

Cryo-Electron Tomography of Whole Cells

Beyond Cryo-ET's application in cellular organelles, researchers are now pushing the boundaries by attempting to visualize entire cells in three dimensions. This ambitious approach aims to capture the full complexity of cellular architecture, including the spatial distribution of macromolecular complexes and organelles within a cell.

Recent breakthroughs in Cryo-ET of whole cells have provided unprecedented insights into cellular processes, such as the organization of the nuclear pore complex and the spatial arrangement of ribosomes in the cytoplasm. These studies have the potential to revolutionize our understanding of cell biology at the molecular level.

Artificial Intelligence and Machine Learning in Structural Biology

The integration of artificial intelligence (AI) and machine learning (ML) techniques is rapidly transforming structural biology. AI/ML algorithms are being used to accelerate data analysis, automate image processing in cryo-EM, predict protein structures, and improve the accuracy of structural models.

For instance, AI-driven algorithms can assist in particle picking and classification in cryo-EM, reducing the time required for data processing. Additionally, AI-powered protein folding predictions have made substantial progress, as showcased by DeepMind's AlphaFold, which can predict protein structures with remarkable accuracy.

As structural biology enters a new era, these emerging techniques offer exciting prospects for researchers. Cryo-ET enables the study of complex cellular processes, in-cell NMR opens doors to in vivo structural investigations, and XFELs promise high-resolution data from previously challenging samples. Moreover, advances in solid-state NMR, cryo-EM, and the integration of AI/ML are poised to drive the field to unprecedented heights. These innovations not only expand our structural biology toolkit but also deepen our understanding of the molecular intricacies of life, offering new possibilities for drug discovery and therapeutic interventions. Structural biology is on the brink of transformative discoveries, and these techniques are at the forefront of this scientific revolution.

18.2 Artificial intelligence and machine learning in structure determination

Looking at the significance of structural biology, one of the most transformative forces in recent years has been the integration of artificial intelligence (AI) and machine learning (ML) into the process of protein structure determination. These technologies have not only accelerated the pace of discovery but have also unlocked novel insights into complex biological systems. In this subsection, we delve into the fascinating realm of AI and ML applications in high-resolution protein structure determination, exploring how these computational tools are reshaping the field.

Unveiling the Power of Deep Learning

Deep Learning in Cryo-EM

Cryo-electron microscopy (cryo-EM) has experienced a renaissance, largely thanks to deep learning techniques. Traditional cryo-EM image analysis required painstaking manual intervention, but deep learning algorithms have revolutionized this process. A standout example is the development of neural networks, such as convolutional neural networks (CNNs), for particle picking and image classification. These networks are capable of distinguishing between signal and noise in raw cryo-EM images with remarkable accuracy. For instance, the DeepPicker algorithm, based on a CNN architecture, has been used to automate particle picking, significantly reducing human intervention and expediting data acquisition. This approach not only enhances efficiency but also reduces the risk of human bias.

Moreover, deep learning has facilitated breakthroughs in 3D reconstruction from cryo-EM images. Single-particle analysis, a cornerstone of cryo-EM, often involves the alignment and classification of thousands of individual particle images to generate a 3D reconstruction. This process, which was once computationally intensive and time-consuming, now benefits from deep learning-powered algorithms. Techniques like cryoSPARC and RELION employ neural networks for particle alignment and classification, enabling researchers to obtain high-resolution structures more rapidly than ever before.

AI-Driven Protein Folding Prediction

Predicting protein structures from amino acid sequences has been a longstanding challenge in structural biology. However, AI and ML have breathed new life into this problem. AlphaFold, a deep learning-based system developed by DeepMind, made headlines with its remarkable ability to predict protein structures with atomic-level accuracy. By training on a vast dataset of known protein structures, AlphaFold can rapidly and accurately predict the 3D structures of proteins, even those with no homologous structures in existing databases.

AlphaFold's impact extends beyond structure prediction; it also aids in experimental design. Researchers can use predicted structures as a guide to design experiments, select suitable crystallization conditions, or choose optimal NMR experiments, streamlining the entire structure determination process.

Enhancing X-ray Crystallography with Machine Learning

Intelligent Data Analysis

X-ray crystallography, a cornerstone of structural biology, has also harnessed the power of machine learning. ML algorithms have proven invaluable in automating the analysis of X-ray diffraction data. For example, the AutoSol program employs ML techniques to determine phases from diffraction data, a critical step in X-ray crystallography. ML models are trained to recognize patterns in electron density maps, expediting the solution of crystal structures.

Predicting Crystallographic Outcomes

Machine learning can help predict the success of crystallization experiments. By analysing historical data on crystallization conditions and outcomes, ML models can identify the conditions most likely to yield high-quality crystals. This predictive capability saves researchers time and resources by guiding them toward conditions with a higher probability of success.

NMR Spectroscopy and Machine Learning Synergy

Spectrum Analysis and Resonance Assignment

Nuclear magnetic resonance (NMR) spectroscopy is a powerful tool for studying protein structures in solution. However, interpreting NMR spectra can be challenging due to signal overlap and noise. Machine learning models can assist in the automated analysis of NMR spectra, aiding in resonance assignment and structure determination. Programs like MARS (Magnetic Resonance Assignment by Sparky) leverage ML to automate the assignment of NMR resonances, reducing the tedious manual effort traditionally required.

Structural Refinement and Validation

ML techniques have found utility in NMR structural refinement and validation. Algorithms can predict NMR-derived structural parameters, such as dihedral angles and NOE distances, and compare them with experimental data to identify structural outliers. This streamlines the process of structure refinement and ensures the accuracy of the final models.

Challenges and Considerations

While AI and ML hold tremendous promise in protein structure determination, they also present challenges and considerations. First and foremost is the need for high-quality training data. ML models rely on large, diverse datasets for training, and the availability of such datasets can be a limiting factor. Additionally, overfitting—a common pitfall in ML—must be carefully addressed to ensure the generalizability of models.

Ethical concerns also arise, particularly regarding the potential for data privacy breaches. Structural biology datasets contain sensitive information about biomolecules, and the application of AI and ML must be conducted with a keen awareness of data security and privacy.

Future Prospects

The marriage of AI and ML with high-resolution protein structure determination has opened new avenues of exploration in structural biology. As these technologies continue to evolve, we can expect even more rapid and accurate structure determination, enabling researchers to delve deeper into the mysteries of biology. Moreover, the synergistic relationship between experimental techniques and computational models promises to unravel complex biological systems with unprecedented precision.

Thus, the integration of artificial intelligence and machine learning into the realm of high-resolution protein structure determination has ushered in a new era of discovery and innovation. These computational tools are not merely accelerants; they are powerful catalysts that are transforming our understanding of the structural intricacies of life itself. The future of structural biology is bright, with AI and ML poised to play an ever-expanding role in unravelling the mysteries of the molecular world.

18.3 Single-particle cryo-EM and beyond

In recent years, single-particle cryo-electron microscopy (cryo-EM) has emerged as a transformative technique in structural biology, enabling the determination of high-resolution structures of biological macromolecules without the need for crystallization. This section explores the principles of single-particle cryo-EM, its applications, and the exciting developments that lie beyond its current capabilities.

Principles of Single-Particle Cryo-EM

Cryo-EM is rooted in the concept of imaging biological specimens at cryogenic temperatures, which reduces radiation damage and minimizes structural distortions due to specimen dehydration. The single-particle approach involves imaging individual, isolated macromolecular complexes embedded in vitreous ice. The key steps in single-particle cryo-EM are as follows:

Sample Preparation: Biological samples, such as purified proteins or complexes, are applied to a thin carbon film grid coated with a layer of amorphous carbon. Excess sample is blotted away, leaving a thin layer of solution covering the grid.

Freezing: The grid is rapidly plunged into liquid ethane or propane, which freezes the sample in a thin layer of vitrified ice. This vitrified ice matrix preserves the native structure of the macromolecules.

Imaging: The grid, now embedded in ice, is transferred to the cryo-EM microscope. Electron micrographs are acquired by directing a beam of high-energy electrons through the sample. The resulting images contain information about the sample's structure.

Data Processing: The acquired images are subjected to extensive computational analysis. Individual particles are selected from the micrographs, aligned, and classified into groups based on their structural similarities.

3D Reconstruction: A 3D density map of the macromolecule is generated by combining information from multiple 2D images of individual particles. This process, known as 3D reconstruction, results in a high-resolution structure of the specimen.

Applications of Single-Particle Cryo-EM

Macromolecular Structures

Single-particle cryo-EM has revolutionized our understanding of the structures of large biological macromolecules, such as ribosomes, viruses, and membrane proteins. For example, the groundbreaking work on the structure of the ribosome by Venkatraman Ramakrishnan, Thomas A. Steitz, and Ada E. Yonath, which earned them the Nobel Prize in Chemistry in 2009, relied heavily on cryo-EM. This technique allows researchers to visualize complex assemblies and dynamic conformations of biomolecules at near-atomic resolution, providing insights into their function.

Drug Discovery and Design

The high-resolution structures obtained through cryo-EM have opened new avenues in drug discovery. Researchers can now visualize the interaction between potential drug compounds and their target proteins with unprecedented detail. This structural information is invaluable for rational drug design and the optimization of drug candidates. For instance, the cryo-EM structure of the SARS-CoV-2 spike protein complexed with neutralizing antibodies has informed the development of therapeutic antibodies for COVID-19.

Structural Heterogeneity

One of the strengths of cryo-EM is its ability to capture structural heterogeneity within a sample. Macromolecular complexes often exist in multiple conformations, and cryo-EM can reveal these variations. This capability has shed light on the flexibility of biomolecules and their functional relevance. For instance, cryo-EM studies of the ATP synthase enzyme have revealed how it adapts to different functional states by undergoing conformational changes.

Challenges and Current Limitations

While single-particle cryo-EM has made remarkable strides, it still faces several challenges and limitations:

Resolution: Although cryo-EM can achieve near-atomic resolution, obtaining atomic-level details for smaller molecules or flexible regions within macromolecules remains challenging.

Sample Size: Cryo-EM is best suited for large macromolecular complexes. Smaller proteins or complexes may require specialized techniques or the combination of cryo-EM with other structural biology methods.

Radiation Damage: Even at cryogenic temperatures, the high-energy electrons used in cryo-EM can cause radiation damage to the sample over time. This limits the total dose of electrons that can be applied to the specimen.

Data Acquisition Speed: Data collection in cryo-EM can be time-consuming, particularly for samples with low contrast or conformational heterogeneity. Faster data acquisition methods are being developed to address this issue.

Future Directions and Beyond Single-Particle Cryo-EM

As the field of cryo-EM continues to evolve, several exciting developments are on the horizon:

High-Frame-Rate Imaging

One of the limitations of cryo-EM is the relatively slow frame rate of detectors, which can limit the study of dynamic processes. Advances in detector technology are expected to enable high-frame-rate imaging, allowing researchers to capture rapid conformational changes in biomolecules.

Correlative Microscopy

Correlative microscopy combines cryo-EM with other imaging techniques, such as super-resolution fluorescence microscopy, to provide complementary information about a sample's structure and function. This integrated approach enhances the interpretability of cryo-EM data.

In Situ Cryo-EM

In situ cryo-EM allows researchers to study biological samples in their native cellular context. This approach promises insights into the spatial organization of macromolecules within cells and tissues, bridging the gap between structural biology and cell biology.

Cryo-Electron Tomography (Cryo-ET)

Cryo-ET extends the capabilities of cryo-EM by reconstructing 3D structures of thick specimens, such as intact cells or organelles. This technique is poised to provide detailed insights into cellular architecture and subcellular processes.

Artificial Intelligence and Machine Learning

The application of artificial intelligence and machine learning algorithms to cryo-EM data analysis is accelerating the process of image processing, particle picking, and 3D reconstruction. These tools are enhancing the efficiency and accuracy of cryo-EM studies.

Single-particle cryo-EM has revolutionized structural biology by enabling the determination of high-resolution structures of challenging macromolecules. Its applications in drug discovery, the study of structural heterogeneity, and our understanding of complex biomolecular assemblies have expanded our knowledge of the molecular world. As technology continues to advance, the future of cryo-EM holds promise for even higher resolution, faster data acquisition, and the integration of cryo-EM with other imaging techniques, further advancing our understanding of the intricate structures that underlie life's processes.

Chapter 19: Ethical and Societal Implications

19.1 Ethical considerations in structural biology research

The pursuit of knowledge through scientific research is an endeavour that holds great promise for humanity. Within the realm of structural biology, the determination of high-resolution protein structures has far-reaching implications, from understanding the molecular basis of diseases to facilitating drug discovery. However, this scientific endeavour is not devoid of ethical considerations that warrant careful attention. In this subsection, we delve into the ethical dimensions of structural biology research, exploring issues related to privacy, dual-use concerns, data sharing, and responsible conduct within the scientific community.

Privacy Concerns in Structural Biology

One of the primary ethical concerns in structural biology research centres around the privacy of individuals whose biological materials are used for experimentation. While much of structural biology relies on the study of purified proteins or macromolecular complexes, some studies require the use of biological samples obtained from living organisms. This may include tissues, cells, or even genetic material.

Example 1: Human Tissue Samples

Consider a scenario where a structural biologist is studying a protein implicated in a rare genetic disorder. To elucidate the protein's structure and function, the researcher requires access to tissue samples from affected individuals. In such cases, maintaining the privacy and confidentiality of the donors is paramount. Researchers must obtain informed consent and adhere to strict ethical guidelines to ensure that the donors' identities and medical histories are protected.

Additionally, data obtained from these samples, such as high-resolution structures, must be de-identified to prevent any potential re-identification of the donors. Failure to protect the privacy of individuals can not only lead to ethical breaches but also hinder future research efforts as potential donors may become reluctant to participate.

Dual-Use Concerns in Structural Biology

Dual-use concerns refer to the potential for scientific knowledge or technologies developed for benevolent purposes to be misused for harmful purposes, such as bioterrorism or the creation of biological weapons. While the majority of structural biology research is conducted for peaceful and beneficial purposes, there is a need to be vigilant about the possibility of unintended consequences.

Example 2: Gain-of-Function Research

In recent years, there has been intense debate surrounding gain-of-function research, which involves making modifications to biological agents to enhance their virulence, transmissibility, or other characteristics. In structural biology, researchers may study viruses or toxins to understand their structures and mechanisms, with the aim of developing vaccines or therapeutics. However, there is a fine line between legitimate research and research that could potentially enable the creation of more dangerous pathogens.

Ethical considerations in this context involve weighing the benefits of research against the potential risks. Researchers must carefully assess the implications of their work and ensure that appropriate safety and security measures are in place to prevent misuse.

Data Sharing and Open Science

The principles of openness and collaboration are fundamental to the scientific enterprise. Structural biology is no exception, as the sharing of data and research findings accelerates progress and fosters innovation. However, ethical dilemmas can arise when it comes to determining what data should be shared, when, and with whom.

Example 3: Data Sharing in Cryo-Electron Microscopy

Cryo-electron microscopy (cryo-EM) has revolutionized structural biology by enabling the determination of high-resolution structures of challenging biological macromolecules. Data sharing in cryo-EM has been a subject of debate. On one hand, open-access initiatives such as the Electron Microscopy Data Bank (EMDB) and Protein Data Bank (PDB) have greatly benefited the scientific community by making structural data freely available. On the other hand, concerns exist regarding the potential misuse of data by malicious actors, particularly in the context of biotechnology.

To address these concerns, responsible data-sharing practices have been developed, including controlled access mechanisms and data deposition policies that balance the need for transparency with security considerations. These practices help strike a balance between the benefits of open science and the need to safeguard sensitive information.

Responsible Conduct in Structural Biology

Beyond specific ethical concerns, structural biologists are also expected to adhere to the broader principles of responsible conduct in research. This entails maintaining the highest standards of integrity, honesty, and transparency in their work.

Example 4: Authorship and Publication Ethics

The process of publishing research findings involves ethical considerations related to authorship and credit. All individuals who make a significant intellectual contribution to a study should be acknowledged as authors. However, disputes over authorship and issues of plagiarism or data fabrication can erode the trust and integrity of the scientific community.

Structural biologists must follow established publication guidelines and ethical standards to ensure that credit is appropriately attributed and that research findings are reported accurately and honestly.

Ethical considerations are an integral part of structural biology research. Researchers in this field must navigate privacy concerns when working with biological samples, remain vigilant about dual-use concerns, and strike a balance between data sharing and security. Moreover, upholding the principles of responsible conduct in research is essential for maintaining the trust of the scientific community and the broader public.

As structural biology continues to advance, ongoing discussions and awareness of ethical issues are critical to ensure that the benefits of this research are realized without compromising ethical principles and societal values. Researchers, institutions, and policymakers must collaborate to develop and implement ethical frameworks that guide the responsible pursuit of knowledge in this exciting field.

19.2 Privacy and security concerns with structural data

In the age of big data and rapid technological advancements, privacy and security have become paramount concerns across various scientific disciplines, including structural biology. While the determination of high-resolution protein structures is undoubtedly a remarkable achievement, it comes with its own set of ethical and security challenges that cannot be ignored. This subsection delves into the multifaceted realm of privacy and security concerns associated with structural data, exploring real-world examples and the potential consequences of mishandling such sensitive information.

The Unveiling of Personal Genetic Information

One of the most profound implications of structural biology lies in its potential to reveal sensitive personal information. Consider the case of a pharmaceutical company conducting research on a novel drug target, a protein implicated in a rare genetic disorder. As part of their structural biology efforts, they successfully determine the high-resolution structure of this protein.

However, when this structural information is made public or inadvertently leaks, it can lead to unintended consequences. For instance, individuals with the rare genetic disorder may be identified based on their genetic sequences, and this could potentially lead to discrimination in health insurance, employment, or even stigmatization within their communities. This scenario underscores the importance of stringent privacy safeguards when dealing with structural data.

Data Sharing and Global Collaboration

Scientific progress thrives on collaboration and data sharing. Structural biologists routinely share their findings in public databases and research publications, enabling others in the field to build upon their work. However, this openness comes with inherent risks.

Consider a scenario where a research institution in one country collaborates with another institution in a different jurisdiction to study a virus's protein structure. During the course of their research, they share structural data, including the sequence and coordinates of the protein, with their collaborators. Inadvertently, this sensitive information could fall under the purview of different privacy laws and regulations, leading to legal complications and potential breaches of confidentiality.

Biotechnology Patents and Intellectual Property

The race to develop new drugs and therapies based on protein structures has led to an increasing number of biotechnology patents. These patents often include detailed structural information about the proteins and their binding sites. While patents are essential for protecting intellectual property, they also raise privacy and security concerns.

For instance, consider a biotechnology company that secures a patent for a novel drug target's structure. This patent grants them exclusive rights to develop therapies targeting this protein. However, the detailed structural information disclosed in the patent may inadvertently reveal potential vulnerabilities in the human body. This information could be exploited by malicious actors with harmful intent, such as designing bioweapons or engaging in bioterrorism.

Data Hacks and Cybersecurity Threats

In the digital age, structural biology data is stored electronically and shared through online platforms. This opens the door to cybersecurity threats and data breaches, which can have far-reaching consequences.

A notorious example is the 2015 cyberattack on the Ashley Madison website, a platform designed for extramarital affairs. Hackers accessed and released user data, exposing the personal information of millions of individuals. While this incident is unrelated to structural biology, it serves as a stark reminder of the potential risks associated with data storage and privacy breaches.

In the context of structural biology, a cyberattack on a research institution's database could result in the theft of sensitive structural data. This stolen information could then be misused for a range of malicious purposes, including identity theft, extortion, or even the creation of harmful biological agents.

Case Study: The PDB Hack

One of the most prominent instances highlighting the security vulnerabilities in structural biology occurred in 2003 when the Protein Data Bank (PDB) experienced a security breach. The PDB is a repository of protein structure data accessible to researchers worldwide. In this breach, hackers gained unauthorized access to the PDB and managed to alter and delete several structure files.

While the motive behind this breach remains unclear, it raised concerns about the security of vital structural data repositories. Had the hackers chosen to manipulate or delete data pertaining to critical biological molecules, the consequences could have been catastrophic. This incident underscores the need for robust cybersecurity measures in safeguarding structural data.

Mitigating Privacy and Security Concerns

The aforementioned examples illustrate the multifaceted nature of privacy and security concerns associated with structural data. However, the field of structural biology is not helpless in the face of these challenges. Several measures can be adopted to mitigate these concerns effectively:

Data Anonymization: Structural biologists should prioritize anonymizing data by removing any personally identifiable information before sharing it. This can help protect the privacy of individuals whose genetic information might be indirectly revealed through structural data.

Ethical Review: Research institutions and journals can implement rigorous ethical review processes for structural biology research. This includes evaluating the potential consequences of publishing sensitive structural information and assessing its implications for privacy and security.

Cybersecurity: Institutions housing structural biology data must invest in robust cybersecurity measures, including encryption, intrusion detection systems, and regular security audits, to protect against data breaches and cyberattacks.

International Collaboration: When engaging in global collaborations, researchers should be mindful of varying privacy regulations and seek legal counsel to ensure compliance with relevant laws.

Public Awareness: It is crucial to raise awareness among researchers, policymakers, and the general public about the potential privacy and security risks associated with structural data. This can lead to more responsible data handling and informed decision-making.

The pursuit of high-resolution protein structures has revolutionized our understanding of biology and opened new avenues for drug discovery and therapeutics. However, as the field continues to advance, it must navigate the intricate terrain of privacy and security concerns. The examples presented in this subsection emphasize the importance of vigilance, ethical considerations, and international cooperation in protecting sensitive structural data. Striking a balance between scientific openness and safeguarding individual privacy and national security is a complex but essential task for the future of structural biology.

19.3 Responsible use of structural information

The elucidation of protein structures has transformed our understanding of molecular biology and revolutionized drug discovery and biotechnology. However, the immense power of this knowledge also comes with ethical and societal responsibilities. In this chapter, we delve into the critical topic of responsible use of structural information, exploring the ethical dilemmas, privacy concerns, and security issues associated with protein structure data.

Ethical Considerations

Access and Distribution of Structural Data

One of the foremost ethical considerations in the realm of protein structure determination is the accessibility and distribution of structural data. Protein structures are often deposited in public databases, such as the Protein Data Bank (PDB), making them accessible to researchers worldwide. This open-access approach fosters scientific collaboration and accelerates discoveries. However, it also raises questions about data ownership, authorship, and control.

Example 1: The Case of Henrietta Lacks

In 1951, without her knowledge or consent, cancer cells were harvested from Henrietta Lacks, an African American woman. These cells, known as HeLa cells, became instrumental in scientific research and led to numerous discoveries, including the development of the polio vaccine. The ethical dilemma of using her cells without informed consent raised questions about ownership and compensation for contributions to scientific knowledge.

Similarly, in structural biology, data generated from biological samples may raise questions about the source's consent and whether they should have a say in the data's usage.

Dual-Use Dilemma

The dual-use dilemma refers to situations where scientific knowledge or technologies developed for beneficial purposes can also be misused for harmful ends, such as bioterrorism or biowarfare. In the context of protein structure determination, researchers often wrestle with the dual-use nature of their work.

Example 2: The 2018 CRISPR-Baby Controversy

Chinese scientist He Jiankui's announcement in 2018 that he had edited the genomes of twin babies using CRISPR-Cas9 technology shocked the world. While CRISPR-Cas9 holds immense promise for treating genetic diseases, its potential for misuse, including creating designer babies with enhanced traits, raised ethical concerns.

In structural biology, the knowledge gained about protein structures can be used for both beneficial purposes, such as drug discovery, and potentially harmful ones, like designing more potent toxins or bioweapons.

Privacy and Security Concerns

Personalized Medicine and Genomic Privacy

Advances in structural biology have enabled the development of personalized medicine, tailoring treatments to an individual's unique genetic makeup. While this holds great promise for improving healthcare outcomes, it also brings privacy concerns regarding the storage and use of genomic data.

Example 3: The Myriad Genetics BRCA Gene Patent

In 2013, the U.S. Supreme Court ruled against Myriad Genetics' patent claims on the BRCA genes, which are linked to breast and ovarian cancer. The decision affirmed that genes occurring naturally in the human body cannot be patented. This case highlighted the ethical issues surrounding genetic data ownership and access, as well as the potential for genetic discrimination.

In structural biology, the genetic information embedded in protein structures can reveal sensitive health-related data, making it crucial to safeguard this information.

Data Security and Cyber Threats

With the increasing digitization of structural data, concerns about data security and cyber threats have become paramount. Malicious actors may attempt to steal valuable structural information or disrupt research efforts.

Example 4: The 2018 Russian GRU Hacking of OPCW

In 2018, the Russian military intelligence agency GRU attempted to hack the Organization for the Prohibition of Chemical Weapons (OPCW) in the Netherlands. The organization was investigating the use of chemical weapons in Syria. This incident underscored the importance of securing sensitive research data.

In structural biology, maintaining the integrity and confidentiality of high-resolution protein structures is essential to prevent misuse or unauthorized access.

Responsible Conduct in Research

Ethical Guidelines and Oversight

To address the ethical challenges in structural biology, various organizations and institutions have developed ethical guidelines and oversight mechanisms. Researchers are expected to adhere to these principles to ensure responsible conduct in their work.

Example 5: The Declaration of Helsinki

The Declaration of Helsinki is a set of ethical guidelines for medical research involving human subjects. It emphasizes the importance of informed consent, risk-benefit assessment, and participant confidentiality. Researchers in structural biology working with human samples or data must consider these principles.

Promoting Ethical Education

Educating researchers and students about the ethical considerations in structural biology is crucial. Universities and research institutions should integrate ethics education into their curricula to foster a culture of responsible research.

Example 6: Responsible Conduct of Research (RCR) Training

Many universities and funding agencies require researchers to undergo Responsible Conduct of Research (RCR) training, which covers topics like data management, authorship, and research integrity. RCR training ensures that scientists are aware of and can navigate the ethical challenges in their work.

High-resolution protein structure determination is a powerful tool with profound implications for science and society. However, the responsible use of structural information is paramount. Researchers, policymakers, and society as a whole must grapple with ethical dilemmas, privacy concerns, and security issues to ensure that this knowledge is harnessed for the greater good while safeguarding against misuse. By adopting ethical guidelines, promoting responsible conduct, and fostering education, we can navigate the complex landscape of structural biology with integrity and responsibility.

Chapter 20: Future Perspectives

20.1 The evolving landscape of protein structure determination

The field of protein structure determination has witnessed remarkable evolution over the decades. In this section, we will delve into the dynamic landscape of structural biology, exploring the driving forces, transformative technologies, and emerging trends that are shaping the future of this field.

Revolution in Cryo-Electron Microscopy (Cryo-EM)

One of the most profound revolutions in protein structure determination has been the resurgence of cryo-electron microscopy (Cryo-EM). Historically used primarily for large and symmetrical complexes, Cryo-EM has undergone a metamorphosis. Recent advancements, such as the development of direct electron detectors and improved data processing algorithms, have propelled this technique into the limelight.

Cryo-EM's impact extends across various domains. For instance, in 2017, the Nobel Prize in Chemistry was awarded to Jacques Dubochet, Joachim Frank, and Richard Henderson for their contributions to Cryo-EM, demonstrating the technique's significance. The 'resolution revolution' in Cryo-EM has allowed scientists to visualize biological macromolecules, including proteins, at near-atomic resolution, unveiling intricacies that were once elusive.

A prime example is the structural elucidation of the TRPV1 ion channel, responsible for the sensation of heat and pain. In 2019, researchers used Cryo-EM to capture the channel's structure in multiple conformations, revealing the dynamic mechanisms underlying its function. These breakthroughs exemplify Cryo-EM's potential to decipher intricate protein architectures and dynamics.

Integrative Structural Biology

As the saying goes, 'the whole is greater than the sum of its parts,' and this holds true in structural biology. Integrative structural biology has emerged as a powerful approach that combines data from multiple sources, such as X-ray crystallography, NMR spectroscopy, and Cryo-EM, to construct comprehensive structural models.

For instance, the elucidation of the ribosome's structure, a fundamental cellular machine, exemplifies integrative structural biology's potency. By integrating data from X-ray crystallography and Cryo-EM, researchers achieved a detailed understanding of the ribosome's structure and function, paving the way for insights into protein synthesis and antibiotic development.

Additionally, the concept of hybrid methods has gained traction. Researchers are combining data from various sources to overcome each technique's limitations. For instance, Cryo-EM and NMR data can be merged to obtain higher resolution structures than what each method can provide individually.

Artificial Intelligence and Machine Learning

Artificial intelligence (AI) and machine learning (ML) are reshaping the landscape of protein structure determination. These computational tools have become indispensable in the analysis of vast structural datasets, accelerating the process and enhancing accuracy.

In the kingdom of Cryo-EM, AI-powered algorithms are automating particle picking, 3D reconstruction, and model refinement. This automation not only reduces the manual workload but also improves the consistency and quality of results. For instance, the DeepEMhancer algorithm, developed in 2020, uses deep learning to enhance the quality of Cryo-EM reconstructions, pushing the resolution limits even further.

Machine learning techniques are also aiding in protein structure prediction. AlphaFold, developed by DeepMind, made headlines by predicting protein structures with remarkable accuracy during the 14th Critical Assessment of Structure Prediction (CASP14) competition. This breakthrough has the potential to accelerate the determination of protein structures, especially for proteins that are challenging to crystallize or study using traditional methods.

Single-Particle Cryo-EM and Beyond

While Cryo-EM has already made significant strides, the future holds even greater promise. Single-particle Cryo-EM, which involves the study of individual particles without the need for crystallization, is on the rise. This technique has the potential to unlock the structures of previously inaccessible proteins, membrane proteins, and dynamic complexes.

For example, in 2021, a team of scientists used single-particle Cryo-EM to reveal the structure of the NADH-Ubiquinone

Oxidoreductase (Complex I), a vital component of the respiratory chain. This breakthrough not only shed light on the complex's architecture but also provided insights into its role in cellular respiration and diseases.

Beyond single-particle Cryo-EM, other emerging techniques, such as MicroED (Microcrystal Electron Diffraction) and Solid-State NMR, are pushing the boundaries of protein structure determination. MicroED, in particular, allows researchers to analyze microcrystals, enabling the study of challenging targets and small molecules.

Structural Biology in Drug Discovery

The evolving landscape of protein structure determination has profound implications for drug discovery. High-resolution structures of drug targets provide valuable insights for rational drug design, enabling the development of more effective and specific therapeutics.

Consider the case of the COVID-19 pandemic. The rapid determination of the SARS-CoV-2 spike protein's structure using Cryo-EM played a pivotal role in the development of vaccines and antiviral drugs. This example highlights the agility and responsiveness of structural biology in addressing urgent global health challenges.

Additionally, structural biology is advancing our understanding of protein-ligand interactions, elucidating the binding mechanisms of small molecules to their targets. This knowledge is invaluable for optimizing drug candidates and minimizing side effects.

Ethical and Societal Implications

As protein structure determination continues to advance, ethical considerations come to the forefront. The sharing of structural data, data privacy, and security concerns must be addressed responsibly. Ensuring that structural information benefits society without compromising individual rights and security is a pressing concern.

Moreover, there is a growing debate about the responsible use of structural data in areas like biotechnology and synthetic biology. The ability to engineer proteins with precision raises questions about biosecurity and the potential misuse of this technology.

Future Perspectives

The future of protein structure determination is teeming with possibilities. As technology continues to advance, we can anticipate even higher resolution structures, faster data acquisition, and broader accessibility to structural biology tools. With AI and ML at the helm, automation will become the norm, making structural biology more efficient and accessible.

Furthermore, as we explore the structures of increasingly complex biological systems, from large macromolecular complexes to dynamic protein assemblies, our understanding of biology's intricacies will deepen. This knowledge will have profound implications for fields ranging from medicine to biotechnology.

The landscape of protein structure determination is undergoing a remarkable transformation. Cryo-EM, integrative structural biology, AI, and emerging techniques are reshaping the field, offering new insights into the molecular world. As ethical considerations accompany these advancements, it is crucial that the structural biology community navigates this evolving landscape with responsibility and foresight, ensuring that the benefits of high-resolution protein structures are maximized for the betterment of society.

20.2 Challenges and opportunities for the field

Protein structure determination has undoubtedly witnessed remarkable progress over the years, with advances in technology and methodologies revolutionizing our understanding of the molecular world. However, as with any scientific discipline, the field faces its fair share of challenges and opportunities. In this section, we will explore some of the pressing issues that researchers grapple with while also shedding light on the exciting prospects that lie ahead.

Challenges in Protein Structure Determination

Membrane Proteins and Large Complexes

One enduring challenge in structural biology is the determination of membrane proteins and large molecular complexes. These proteins play crucial roles in cellular processes and as drug targets, but their structural analysis remains technically demanding due to their size, complexity, and the need to mimic their native environment. For example, G protein-coupled receptors (GPCRs), vital drug targets, are notoriously challenging to crystallize.

Dynamic Structures

Many proteins are dynamic in nature, adopting multiple conformations to perform their functions. Capturing these dynamics in high-resolution structures is challenging. NMR spectroscopy is valuable for studying dynamics, but it may not always provide atomic-level details for all protein regions.

Data Quality and Completeness

The quality and completeness of experimental data are paramount for accurate structure determination. In X-ray crystallography and cryo-EM, obtaining high-resolution data can be hindered by factors like radiation damage, crystal imperfections, or particle heterogeneity. Dealing with imperfect data can compromise the accuracy of the resulting structures.

Sample Preparation and Expression

The process of expressing and purifying proteins can be time-consuming and may lead to misfolding or aggregation. For some proteins, obtaining sufficient quantities for analysis is challenging. Techniques like cell-free protein expression and in-cell NMR are emerging to address these issues.

Modelling of Disordered Regions

Proteins often contain disordered regions that lack well-defined structures. Integrating these regions into structural models is a persistent challenge. Cryo-EM and NMR can provide insights into the flexibility of these regions, but modelling them accurately remains a computational hurdle.

Opportunities in Protein Structure Determination

Cryo-Electron Tomography (Cryo-ET)

Cryo-ET allows the study of biological specimens in their native state, providing three-dimensional structural information at nanometre-scale resolution. This technique is invaluable for understanding cellular structures, organelles, and the spatial organization of macromolecular complexes.

Artificial Intelligence (AI) and Machine Learning:

AI and machine learning are transforming structural biology. These technologies enhance image processing, speed up data analysis, and aid in predicting protein structures. AlphaFold, developed by DeepMind, is a remarkable example of AI's potential in predicting protein structures with high accuracy.

Single-Particle Cryo-EM Advances

Single-particle cryo-EM has evolved rapidly. Advanced detectors and data processing algorithms have improved the achievable resolution. This technique is becoming increasingly accessible to structural biologists, opening doors to a wider range of research areas.

Hybrid Methods and Integrative Structural Biology

Combining data from multiple techniques, such as X-ray crystallography, NMR, and cryo-EM, offers a holistic view of protein structures. Integrative structural biology approaches are becoming more prevalent, allowing researchers to address complex biological questions.

Structural Biology in Drug Discovery

The integration of structural biology into drug discovery has immense potential. Structural information on drug targets can accelerate the development of novel therapeutics. For instance, the design of drugs targeting the spike protein of SARS-CoV-2 was aided by high-resolution structures.

Emerging Techniques and Tools

Solid-State NMR

Solid-state NMR spectroscopy is advancing our understanding of membrane proteins and amyloid fibrils, which are implicated in neurodegenerative diseases. It can provide atomic-level details of protein structures in native-like environments.

Miniaturized Cryo-EM

Miniaturized cryo-EM instruments are making it easier to perform structural studies in smaller laboratories. These instruments are more user-friendly, require less sample, and can be used for routine structural biology research.

In-cell NMR

In-cell NMR spectroscopy allows the study of proteins within living cells. This technique provides insights into how proteins behave in their natural environment, shedding light on their functions and interactions.

Cryo-EM Tomography of Cells

Cryo-EM tomography of cells is expanding our knowledge of cellular organization. It enables the visualization of entire cells and their organelles, providing a more comprehensive understanding of cell biology.

Interdisciplinary Collaboration

One of the most promising avenues for overcoming challenges and advancing the field is interdisciplinary collaboration. Structural biologists are increasingly partnering with experts in computational biology, data science, and chemistry. This synergy enables the development of novel methods, tools, and approaches that are essential for addressing complex biological questions.

Ethical Considerations

As the field of protein structure determination continues to advance, ethical considerations become more prominent. The availability of high-resolution structural information raises questions about data sharing, privacy, and security. Researchers must grapple with the responsible use of structural data to ensure that it benefits society without compromising individual rights and security.

High-resolution protein structure determination is at an exciting juncture. While challenges persist, innovative technologies, interdisciplinary collaboration, and ethical considerations are shaping the future of the field. As we continue to unlock the mysteries of protein structures, we move closer to a deeper understanding of life's fundamental processes and the potential for groundbreaking discoveries in health, medicine, and biotechnology. The challenges are real, but so are the opportunities, making this an exhilarating time for structural biology.

20.3 Predictions for the future of high-resolution structural biology

As we stand at the precipice of a new era in structural biology, it is imperative to gaze into the crystal ball of science and make educated predictions about the future of high-resolution structural biology. Over the decades, this field has witnessed remarkable advancements, from the pioneering days of X-ray crystallography to the recent revolution in cryo-electron microscopy (cryo-EM). In this subsection, we will explore the exciting prospects and potential directions that high-resolution structural biology might take in the coming years.

Cryo-EM Will Continue to Flourish

Cryo-EM has undergone a seismic transformation in recent years, allowing scientists to explore previously uncharted territories in structural biology. We can anticipate that this technique will continue to flourish, driven by innovations in detector technology, sample preparation methods, and data processing algorithms. With the development of higher-resolution direct electron detectors and more efficient data acquisition strategies, the resolution of cryo-EM structures will likely continue to improve. It is not far-fetched to predict that near-atomic or even atomic-level resolution for a broader range of biological macromolecules will become the norm rather than the exception.

A prime example of this trend is the resolution revolution in cryo-EM. In 2017, the Nobel Prize in Chemistry was awarded to Jacques Dubochet, Joachim Frank, and Richard Henderson for their pioneering work in cryo-EM, highlighting the profound impact of this technique on structural biology. With the relentless pursuit of higher resolution, cryo-EM is poised to unlock the structures of challenging specimens, such as flexible complexes and heterogeneous assemblies, providing unprecedented insights into cellular processes.

Integrative Structural Biology Will Prevail

The integration of multiple structural techniques, often referred to as integrative structural biology, will become increasingly prevalent. Scientists will leverage the strengths of X-ray crystallography, NMR spectroscopy, and cryo-EM to tackle complex biological questions that cannot be addressed by a single method alone. Integrative approaches will be used not only to validate structural models but also to provide dynamic, spatial, and functional information at high resolution.

One illustrative example is the elucidation of the mammalian 26S proteasome structure. By combining cryo-EM and X-ray crystallography data, researchers were able to determine the complete atomic structure of this massive molecular machine, shedding light on its dynamic functioning in cellular protein degradation.

Advances in Artificial Intelligence

Artificial intelligence (AI) and machine learning will play an increasingly pivotal role in high-resolution structural biology. AI-powered algorithms will aid in the automation of data collection and processing, significantly reducing the time and effort required for structure determination. For instance, machine learning models can assist in particle picking and improving image contrast in cryo-EM data, thus accelerating structure determination pipelines.

Furthermore, AI-driven drug discovery efforts will benefit from high-resolution structures. Predictive algorithms will enable the rapid screening of potential drug candidates against known protein structures, expediting the drug development process. As an example, AlphaFold, developed by DeepMind, demonstrated the power of AI in predicting protein structures with remarkable accuracy. In the future, similar AI approaches will enhance our ability to understand and manipulate protein structures for therapeutic purposes.

Structural Biology in the Cellular Context

The next frontier in structural biology will involve studying biomolecules within their native cellular context. Researchers will aim to visualize the spatial organization of macromolecules and their interactions within cells, tissues, and even whole organisms. This ambitious goal will be pursued through a combination of cryo-EM tomography, super-resolution microscopy, and other emerging techniques.

One exciting area of development is cellular cryo-EM, where scientists are working towards imaging macromolecular complexes in their native cellular environment. Recent achievements, such as visualizing the structure of the nuclear pore complex within intact cells, exemplify the potential of this approach. In the coming years, we can anticipate breakthroughs that will unveil the structural underpinnings of complex cellular processes.

Personalized Structural Medicine

High-resolution structural biology will have a profound impact on personalized medicine. As our understanding of protein structures and their interactions with drugs advances, clinicians will be better equipped to tailor treatments to individual patients based on their genetic and molecular profiles. This personalized approach holds the promise of improved therapeutic outcomes with fewer side effects.

For instance, the determination of high-resolution structures of drug targets in complex with individual patient variants will enable the design of precision therapies. This approach has already shown promise in cancer treatment, where drugs are selected based on the specific genetic mutations present in a patient's tumour.

Structural Proteomics and Beyond

Structural proteomics initiatives will continue to expand, aiming to determine the structures of entire proteomes. These efforts will provide a comprehensive structural atlas of the protein universe, facilitating the functional annotation of genes and the discovery of novel drug targets.

One notable example is the Structural Genomics Consortium (SGC), which has contributed extensively to the open-access structural biology community. The SGC's focus on understudied protein targets and epigenetic regulators exemplifies the potential of structural proteomics to uncover new biology and therapeutic avenues.

Ethical Considerations and Data Security

As high-resolution structural biology advances, ethical considerations will become increasingly important. Researchers must grapple with questions surrounding data privacy, security, and the responsible use of structural information. Ensuring that structural data remains accessible to the scientific community while safeguarding against potential misuse will be a complex but vital challenge.

In conclusion, high-resolution structural biology stands at the threshold of an exciting future. Cryo-EM will continue to redefine our understanding of biomolecular structures, integrative approaches will provide a holistic view of complex systems, AI will accelerate discovery, and the field will extend into the realm of cellular context and personalized medicine. As we navigate these uncharted waters, it is crucial to maintain ethical vigilance and consider the broader societal implications of our discoveries. The journey ahead promises to be exhilarating, with each new structure revealing the intricate beauty of the molecular world and offering new opportunities for scientific exploration and medical breakthroughs.

www.ingramcontent.com/pod-product-compliance
Lightning Source LLC
Chambersburg PA
CBHW072357290526
45794CB00001B/92